U0163069

东京制造

MADE IN TOKYO

MOMOYO KAIJIMA　JUNZO KURODA　YOSHIHARU TSUKAMOTO

北京联合出版公司
Beijing United Publishing Co.,Ltd.

メイド・イン・トーキョー

[日]贝岛桃代、黑田润三、塚本由晴 著　林煌 译

MOMOYO KAIJIMA
JUNZO KURODA
YOSHIHARU TSUKAMOTO

MADE IN

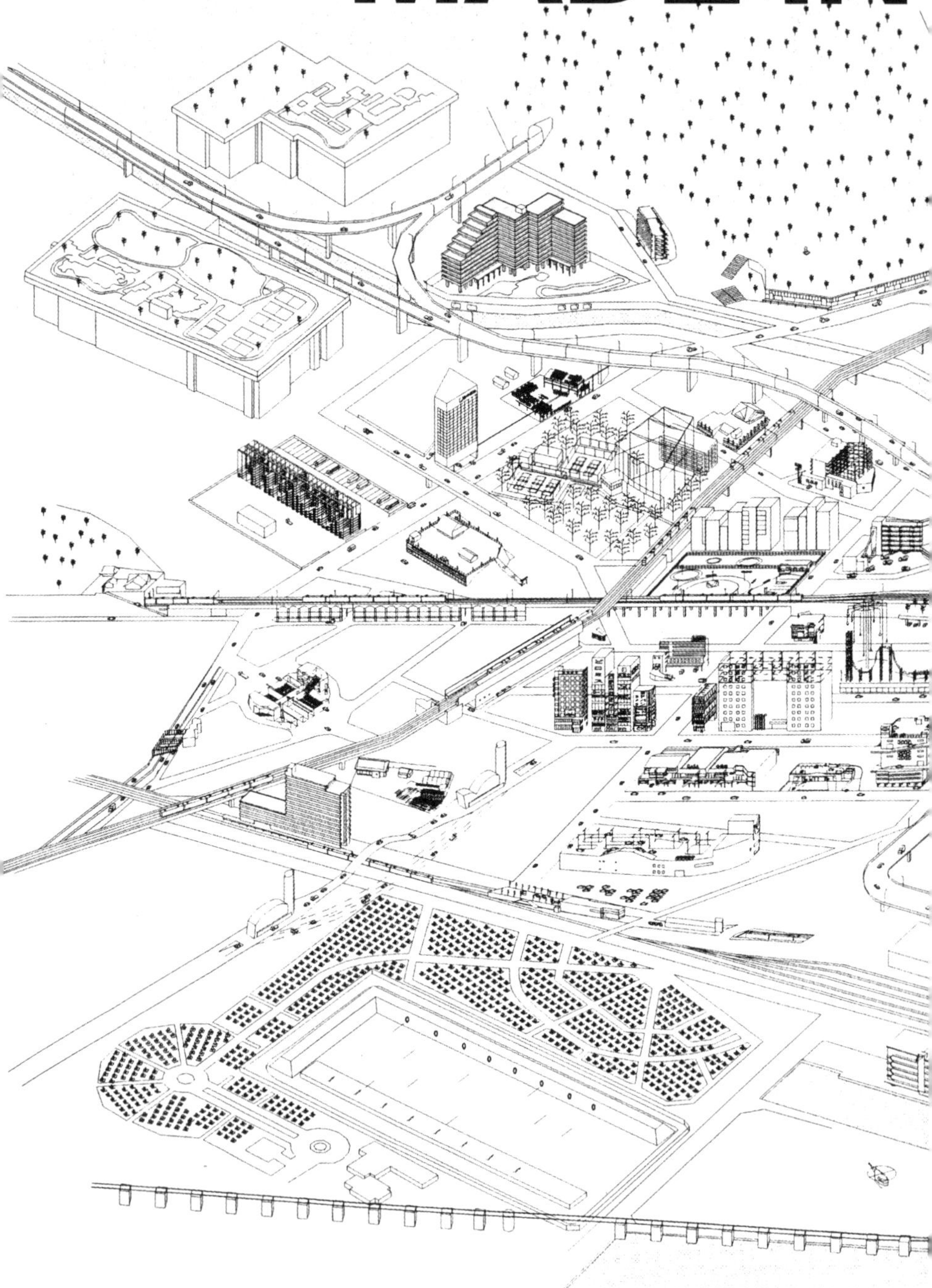

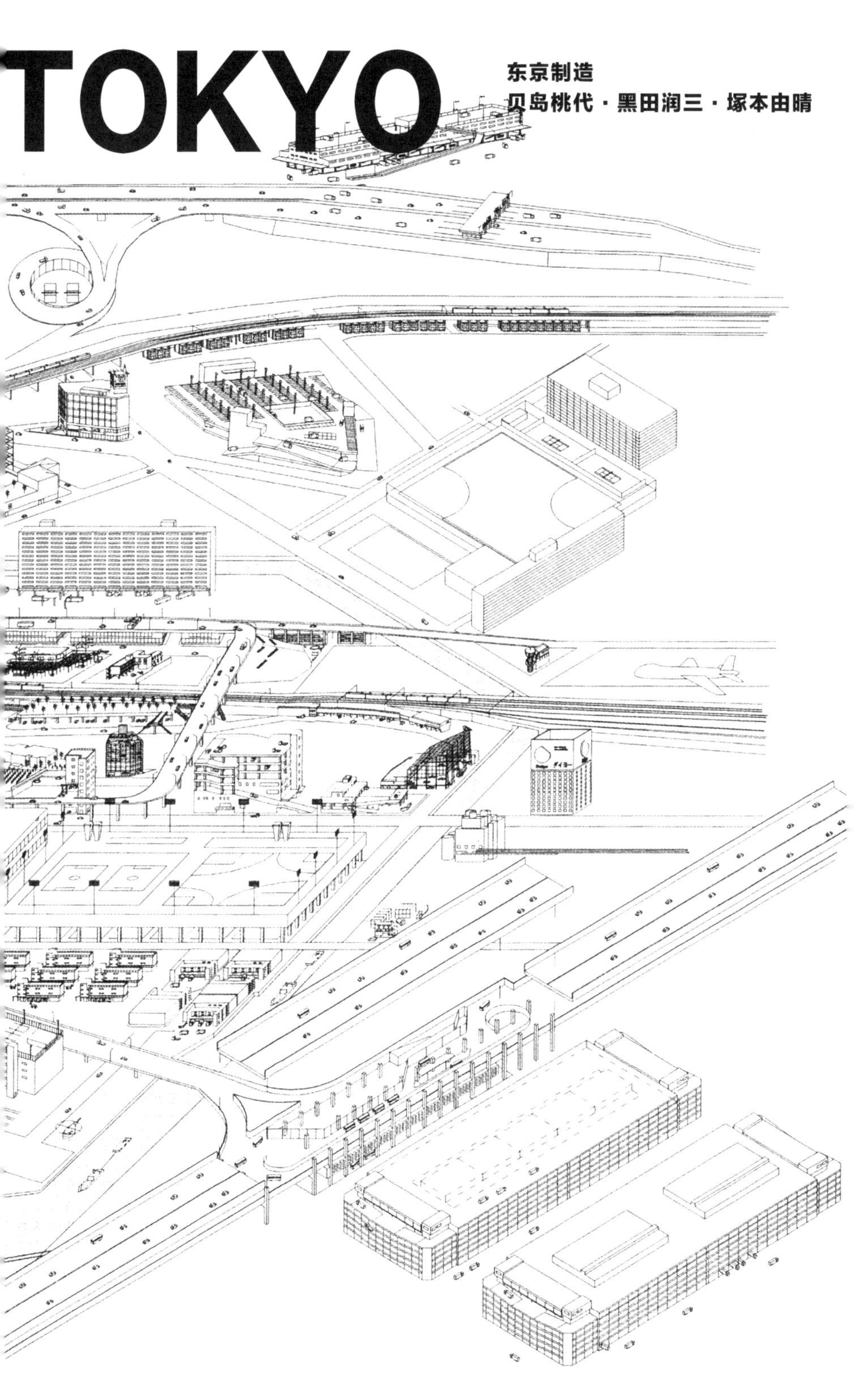

TOKYO

东京制造

贝岛桃代 · 黑田润三 · 塚本由晴

CONTENTS
目录

吸血公园
vampire park

起重机架
crane shelves

幽灵铁道工厂
ghost rail factory

挡土墙公寓
retaining wall apartments

桥洞里的家
bridge home

宅地农场
residential farm

物流立交枢纽
dispersal terminal

皇家高尔夫公寓
royal golf apartments

汽车办公楼
car parking office

绿色停车场
green parking

汽车百货商店
auto department store

家庭餐厅三兄弟
family restaurant triplets

果蔬小镇
vegetable town

松茸形车站楼
sprouting building

TRC（东京物流中心）
tokyo dispersal centre

冷冻园区建筑群
coolroom estate

宠物建筑1号
pet architecture 001

水坝公寓
dam housing

立交枢纽航站楼
airport junction

运动桥
sports bridge

运动者动物园
sportsman zoo

直升机仓库
heli-warehouse

洗车露台
carwash terrace

射击墓地
shooting graveyard

东京制造团队
TEAM MADE IN TOKYO

2001
贝岛桃代 Momoyo Kaijima
黑田润三 Junzo Kuroda
塚本由晴 Yoshiharu Tsukamoto
久野靖广 Yasuhiro Kuno
高木 俊 Shun Takagi
玛丽卡 · 诺伊斯图普尼 Marika Neustupny

nmp 2000
(Web site English version)
贝岛桃代 Momoyo Kaijima
玛丽卡 · 诺伊斯图普尼 Marika Neustupny
黑田润三 Junzo Kuroda
安野 彰 Akira Yasuno
高木 俊 Shun Takagi
塚本由晴 Yoshiharu Tsukamoto
春木裕美子 Yumiko Haruki
荻原富雄 Tomio Ogiwara
汤姆 · 文森特 Tom Vincent
橘 淳子 Junko Tachibana
安部直树 Naoki Abe
饭尾次郎 Jiro Iio
境 洋人 Hiroto Sakai
佐野惠津子 Etsuko Sano

nmp 98
(Web site Japanese version)
贝岛桃代 Momoyo Kaijima
黑田润三 Junzo Kuroda
安野 彰 Akira Yasuno
高木 俊 Shun Takagi
塚本由晴 Yoshiharu Tsukamoto
春木裕美子 Yumiko Haruki
荻原富雄 Tomio Ogiwara
汤姆 · 文森特 Tom Vincent
橘 淳子 Junko Tachibana
安部直树 Naoki Abe
饭尾次郎 Jiro Iio
境 洋人 Hiroto Sakai
佐野惠津子 Etsuko Sano

'96 Architecture
of the Year
贝岛桃代 Momoyo Kaijima
黑田润三 Junzo Kuroda
藤冈 务 Tsutomu Fujioka
久野靖广 Yasuhiro Kuno
安森亮雄 Akio Yasumori
大内靖志 Yasushi Ouchi
柳 博通 Hiromichi Yanagi
荒木裕纪子 Yukiko Araki
盐见理绘 Rie Shiomi
帕潘姐妹 Les Sœurs Papin
克里斯托夫 · 查尔斯 Christoph Charles
塚本由晴 Yoshiharu Tsukamoto

观察协助
correspondents

28 田口阳子 Yoko Taguchi
31 盐见理绘 Rie Shiomi
32 中谷 豪 Go Nakatani
33 小川一人 Kazuto Ogawa
34 米津正臣 Masaomi Yonezu
35 五岛 崇 Takeshi Goshima
36 山口由美子 Yumiko Yamaguchi
37 谷川大辅 Daisuke Tanigawa
38 笠井誉仁 Takanori Kasai
39 小川贵司 Takashi Ogawa
40 小野田 环 Tamaki Onoda
41 浦桥信一郎 Shinichiro Urahashi
43 大成优子 Yuko Oonari
44 川上正伦 Masamichi Kawakami
45 长谷川 豪 Go Hasegwa
46 长冈大树 Daiju Nagaoka
50 中谷礼人 Rhenin Nakatani

中文版编者说明

原书为日 / 英双语对照，日文部分与英文部分的内容并非严格对应。本中文版自日文译出，保留原书英文部分。

INTRODUCTION

东京制造——东京的城市空间和建筑

违和感的出现和消失

从国外尤其是欧洲旅行回到东京时，我总惊讶于大街上竟然有这么多让我觉得违和的地方：铁轨和道路在建筑物的上方穿行，高速公路在河川上蜿蜒展开，汽车沿着坡道一直开到六层建筑的楼顶，住宅区里建起巨大的高尔夫球练习场……在欧洲的主要城市，映入眼帘的仍然是几个世纪前的建筑，从更新换代的角度看，它们还没有实现现代化。与之相比，东京的大部分建筑都是在这三四十年间利用近代之后的技术新造的。这些新技术把寡廉鲜耻的功能组合和空间构成带进了城市。任凭这一切发生的东京，究竟是一个怎样的地方呢？运用相似的建筑技术，东京的建筑又是如何与欧洲的现代主义分道扬镳的呢？

不过，让我感到不可思议的是，仅仅住了不到一周，所有违和感就通通消失不见了，不觉得有些可怕吗？

把身边的事物变成资源

回到从事建筑设计和教学的日常生活之后，我发现无论是建筑杂志，还是大学里使用的教材，收录的都是古今东西的著名建筑。建筑师、评论家和其他专家也都试图从国外的案例和日本的经典作品里找到建筑的典范。虽然这种做法正确且必要，但有一点令我在意：按照这样的价值观，东京会被认为是一个充斥着令人厌恶、毫无价值的建筑的城市。如果你身边只有这样糟糕的城市风景，那么从建筑杰作里寻找建筑典范的做法无非是在标榜自己的品味，而精致的建筑摄影只是在增加多余的期待。在这种情况下，建筑设计好像忽然间变成了一件徒劳的无聊之事，再做下去还有什么意义呢？但是，既然东京已经布满了这样的建筑，我们就应该直面现实，想想如何挖掘它们的优点。如果不能把身边的事物变成丰富的资源，那我们也没必要特意待在东京了。想到这儿，我们不妨开始思考如何变违和为有用，把前面提到的寡廉鲜耻的建筑物视作一份错综复杂的城市现状报告，用于解读东京这座动不动就被贴上“混沌”“混乱”等标签的大都会。

The Appearance and Disappearance of Shamelessness

I'm often surprised when returning to Tokyo, especially, when returning from Europe. Roads and trainlines run over buildings, expressways wind themselves over rivers, cars can drive up ramps to the rooftop of a 6 storey building, the huge volume of a golf practice net billows over a tiny residential district. Most major cities of Europe are still using buildings from previous centuries, and are not modernised in terms of renewing actual building stock. By comparison, almost all buildings in Tokyo have been built within the last 30 or 40 years, utilising contemporary technologies. These technologies have formed a background to the appearance of shameless spatial compositions and functional combinations, unthinkable in the traditional European city. What is it about this city of Tokyo, which can allow such unthinkable productions? How have we managed to arrive at such a different place to European modernity despite being equipped with the same building technology? But one week later, these sorts of questions disappear from my mind, together with the feeling that something is wrong.

Changing our Surroundings into Resources

If we return to our everyday architectural life, architectural magazines and university textbooks are filled with famous works-east and west, old and new. Specialists, such as practitioners and critics find their criteria by looking at overseas examples and Japanese classics. This is correct and necessary, but the values woven by this situation judge this city as consumed by disgusting buildings. But, if our footsteps are actually embedded in such a pitiful urban landscape, the idea of using famous architecture as a criteria base seems to be just an attempt to express good taste. Photographic books amplify a desire for an architecture which simply can't be found in our surroundings. In such a situation, then suddenly architectural design holds no interest anymore; the future appears depressing. If we can't try to turn 'disgusting' buildings into resources, then there is no reason to particularly stay in Tokyo. Surely we can start to think about how to take advantage of them, rather than trying to run away. Shamelessness can become useful. So let's start by considering that these shameless buildings are not collapsible into the concept of 'chaos', but are in fact an intricate reporting of the concrete urban situation.

开始探索

1991年，我们发现了一家狭长的意大利面餐厅，它被塞进了一座建在陡峭斜坡上的棒球练习场下方的缝隙里。棒球练习场和意大利面餐厅在东京随处可见，但这样的组合太不合常理。我们不觉得把练习场建在这里是为了让打球的客人以另一头的旅馆为靶子，或者方便人们在球场挥汗如雨后径直到餐厅饱餐一顿。看着这个不知道该定义为构筑物还是建筑物的组合，我们既怀疑这是个闹着玩的恶作剧，又隐隐对其中蕴含的趣味抱有期待。让人产生这样复杂情绪的地方正是东京。为了不让这种感觉消失，像是初次造访外国的城市那样，我按下快门，把它定格在了照片上。从此，我们开始了探索东京无名的奇妙建筑，也就是"MADE IN TOKYO"(东京制造)的旅程。

失格建筑

结果，最吸引我们的是那些不被建筑美学和形式上的条条框框束缚，执着地优先考虑周边环境和实际需求的建筑。我们既喜爱又嫌弃地称这些建筑物为"失格建筑"。它们大都默默无闻，也不美观，所以在如今的建筑文化里得不到好评，相反还会被当成反面教材。但仔细观察就会发现它们有一个优点：如果希望通过建筑来了解东京的现实，它们比任何建筑师的作品都更合适。如果有所谓艺术的建筑与科学的建筑之分，"失格建筑"两者都不是，勉强可被称为"事件的建筑"吧。这些建筑没有被贴上"东京"的标签，而是反过来展现了东京真实的样子。因此，如果把这些建筑收集、串联起来，东京城市空间的特质不就浮现出来了吗？那段时间正好有一本满是建筑师作品的东京建筑导览手册成了畅销书，但这样的书里没有我们亲身感受到的那个东京。那本导览手册没能回答这样一个问题：我们所在的地方到底拥有怎样的潜力？

与"失格建筑"为邻思考和设计建筑，究竟是怎样的一番体验呢？

Survey Beginnings

In 1991, we discovered a narrow spaghetti shop wrenched into the space under a baseball batting centre hanging from a steep incline. Neither spaghetti shop nor batting centre are unusual in Tokyo, but the packaging of the two together cannot be explained rationally. Despite an apparent convenience in their unity, there is no necessity to hit baseballs towards the opposite hotel, sweat, and then eat at a spaghetti shop. In addition, it is difficult to judge whether this combination is a kind of amusement machine, or a strange architecture. This building simultaneously invited a feeling of suspicion that it was pure nonsense, and expectation in its joyful and willful energy. But we also felt how 'very Tokyo' are those buildings which accompany this ambiguous feeling. Having been struck by how interesting they are, we set out to photograph them, just as though we were visiting a foreign city for the first time. This is the beginning of 'Made in Tokyo', a survey of nameless and strange buildings of this city.

Da-me Architecture

The buildings we were attracted to were ones giving a priority to stubborn honesty in response to their surroundings and programmatic requirements, without insisting on architectural aesthetic and form. We decided to call them 'Da-me Architecture' (no-good architecture), with all our love and disdain. Most of them are anonymous buildings, not beautiful, and not accepted in architectural culture to date. In fact, they are the sort of building which has been regarded as exactly what architecture should not become. But in terms of observing the reality of Tokyo through building form, they seem to us to be better than anything designed by architects. We thought that although these buildings are not explained by the city of Tokyo, they do explain what Tokyo is. So, by collecting and aligning them, the nature of Tokyo's urban space might become apparent. At that time there was a best selling guide book of Tokyo full of architect designed works, but it did not show the bare Tokyo which we felt. It couldn't answer the question of what kind of potentials are in this place we are standing in? What can it mean to think about and design architecture which must stand beside da-me architecture?

在同一个平面上

这场探索基于这样一个假设：每座城市都有能直接反映城市状况和价值观的独特的建筑。就东京而言，我们认为“失格建筑”里藏有思考这座城市及其建筑的线索。但是，“失格建筑”的定义从一开始就不甚清晰，我们一边搜集实例，一边反复讨论定义的问题。在这个过程中注意到的是，不能在没有具体对象的情况下仅仅依靠观念中的模型定义一座城市。20 世纪 80 年代，在“混沌肯定论”和“东京论”的时代背景下，出现了一批建筑师的作品，以空间形式隐喻混乱的城市景观，我们完全不赞同这种借隐喻来概括城市特点的做法。从一开始，我们就有意避开混合多种元素的折中主义、前现代和超现代建筑的比较等带有固定印象的建筑案例。虽然我们赞同路上观察学会从碎片欣赏城市并享受其中的做法，但仍不禁对其中的朴素和怀旧抱有反感，同时我们也不想讨论感伤情绪。真正俘获我们的是那些让人不知如何评价、如何归类的建筑。因此，我们暂且把建筑要素之间的关系视为主要课题，尝试抛开原先加诸建筑的评价和分类。放弃贵族的 / 大众的、美 / 丑、善 / 恶、建筑 / 土木之类的对比，把所有元素放到一个平面上。这样的观察方法正适合东京的城市空间，一个集合了多种多样的结构体的巨大集块（agglomeration）。

无论我们是带着绝望的心情把“混沌”“混乱”这样的标签贴在这个集块上，还是煞有介事地用一个看似合理的故事包装它，都会令我们亲身感受到的那个东京逐渐消失。另外，“每一栋建筑的构成都是由城市空间的结构决定的，可以从中推导出来”，这种说法对于东京而言例外太多，没有任何说服力。因此，如果使用观察城市建筑的常用手段——类型学，则很可能剥夺“失格建筑”中不同元素有趣的混杂性质。那么，该如何观察是好?

导览手册

无论是用故事包装，还是用类型学开道，观察的结果都要以某种形式来展示。请注意，观察的结果取决于展示的方法，如果方法选得

Flatness

The starting hypothesis for the survey is that in any city, the situation and value system of that city should be directly reflected through unique buildings. In the case of Tokyo, we suspect that da-me architecture contains hints to think about the city and architecture. However, the definition of da-me architecture was not necessarily clear from the beginning. We debated at length over each example as we collected them. During these debates, we took care to not think about the city as a conceptual model. In the 1980s there was a background of chaos affirming theory and Tokyology, and the spatial expression of architectural works displayed confusing urban landscape as a metaphor. We strongly wanted to get away from the attitude that the city can be summarised by metaphorical expression. Then again, from the very start, we avoided considering examples which can be read as stereo-typical images such as stylistic eclecticism and contrast between pre- and super-modern. Although we agreed with the Institute of Street Observation's emphasis on pleasure, we felt uncomfortable with the importance attached to modesty and wistfulness. We decided to try to not work with nostalgia. The examples we stuck with were based more on particularity in the way they related directly to use. By treating the relation between elements as the major issue, we tried to see the object without pre-conditioned meanings and categories. We tried to look at everything flatly, by eliminating the divisions between high and low cultures, beauty and ugliness, good and bad. We thought that such a way of seeing is called for by the urban space of Tokyo, which is a gigantic agglomeration of an endless variety of physical structures. If we describe this agglomeration simply as confused or chaotic, or understand it with a predetermined story, then probably our own experience of Tokyo's atmosphere will disperse. Anyway, there are too many exceptions to be able to convincingly deduce each building's composition from the urban structure. So if we try to collapse da-me architecture into a typology, we will lose the interesting mongrel nature of the differing elements. Our flatness means something more specific.

Guidebook

The result of the observation also depends on the method of representation. If the method doesn't suit the observation, the result often can't be grasped. Therefore it is important to develop a method of representation which doesn't lose observational

不好，就会错失非常重要的观察结果。所以，必须要想出一种不会损失观察内容的方法。

我们最后选择的形式是导览手册。东京是一座没有轴线、界限模糊的城市，就像一个不带内置导航系统的巨大迷宫。因此，市面上出现了各种主题的导览手册，东京这座城市被重新编辑以满足不同的需求。尽管导览手册就像事后制作的软件，但它组织了城市的使用方法，所以仍然可以被用于二次规划和个性化定制。而且，导览手册不需要达成什么结论，没有明确的起点和终点，这样的特点正适合东京这座一直在建设和拆除的循环中更新的城市。

建筑师的城市论

在这番探索的过程中，我们从前辈建筑师的建筑理论与城市理论里获益良多。从伯纳德·鲁道夫斯基的《没有建筑师的建筑》里，我们学着通过民间风格的建筑观察周边环境；从尼古拉斯·佩夫斯纳的《建筑类型的历史》里，我们看到在思考建筑史和建筑理论时，他如何在作为分析材料的建筑类型的选择上表现出鲜明的个人主张和批评精神；从阿尔多·罗西的《城市建筑学》里，我们认识到城市的记忆正是由其中大大小小的建筑创造的，建筑和城市是相互依存的关系；从科林·罗和罗伯特·斯拉茨基的《透明性》里，我们发现空间的成立是异质的范畴交错叠加的结果；从罗伯特·文丘里、丹尼斯·斯科特·布朗、史蒂文·艾泽努尔的《向拉斯维加斯学习》里，我们看到了在建筑的历史长河里为"失格建筑"寻找位置的努力；从雷姆·库哈斯的《癫狂的纽约》里，我们感受到城市居民不断膨胀的欲望，以及这些欲望创造的一系列事件，造就了一幅鲜活的当代城市画像；从今和次郎的《考现学入门》里，我们感受到一种观察城市的热情，即原原本本地记录看到的东西，对无论多么不起眼的事物都投以热切的目光；藤森照信等人的《路上观察学》向我们展示了从实际见到的街景碎片出发，发挥想象力，讲述城市的小小历史的乐趣。以上这些论述都让我们相信，虽然观察的是某座特定城市及其建筑，但最后

quality.
The format we chose was that of a guidebook. Tokyo is a giant maze-like city without physical navigational aids such as axes or urban boundary. Perhaps because of this, there are innumerable guidebooks on every facet of life in this city. Tokyo has already been edited to suit every possible objective. Even if they form a kind of software after the fact, in terms of organising the way the city is used, guidebooks can become a tool for urban planning. However, a guidebook doesn't need a conclusion, clear beginning or order. This seems suitable for Tokyo, where the scene is of never ending construction and destruction.

Urban Theory by Architects

Much was learnt from architectural and urban theories from our predecessors. From Bernard Rudofsky's 'Architecture Without Architects', we looked at the response between architecture and the environment in vernacular buildings. From Nikolaus Pevsner's 'A History of Building Types', we considered how he picked up arbitrariness and criticism in the selection of building types as material for thinking about architecture. From Aldo Rossi's 'Architecture of the City', we thought about the interdependent relationship between architecture and the city. From Colin Rowe and Robert Slutzky's 'Transparency: Literal and Phenomenal', we learnt about how space evolves out of the overlapping of various design criteria. From Robert Venturi, Denise Scott Brown and Steven Izenour's 'Learning From Las Vegas' we realised the power of placing 'bad architecture' within the line of architectural history. From Rem Koolhaas's 'Delirious New York', we delighted in the idea that the whole of the contemporary city is made up of a series of accidents, in accordance with inevitable changes to the overall urban plan. From Wajiro Kon's 'Kogengaku Nyumon' (Introduction to Cultural Studies), we gained a love of observing the city before us, and an understanding where even the most subtle things start to hold meaning, sketch by sketch. From Terunobu Fujimori et al's, 'Institute of Street Observation' we discovered the joy of actually walking in the street and finding fragments – allowing the swelling of imaginations and the speaking of small urban histories. We were encouraged to think that each of these theories had been born out of discussing particular cities and architectures. They have concrete origins in a specific place, and yet in the end they lead towards an abstract level, which can open new architectural and urban awareness. What kind of awareness will

能得到的是对于“建筑”“城市”这些概念的全新认识。东京制造的建筑物会带给我们哪些新的认识呢？

从“建筑”到“建筑物”

正如“失格建筑”这个称呼所示，东京制造的建筑物客观地说不怎么美观，也无法用作建筑规划的优秀范本。从建筑类型划分，“失格建筑”里几乎没有图书馆或美术馆这样的A级文化设施，而多为停车场、驾校、棒球练习场这样的B级设施，其中也有不少含有土木构筑物的复合型设施。而且它们都不是知名建筑师的作品。这些“一无是处”的建筑之所以仍然值得关注，是因为它们既不拐弯抹角，也不装模作样，尽可能利用已有的材料，通过务实的方式满足必要的建筑功能。它们不从迂回曲折的历史脉络中寻求文化层面的解答，而是直奔主题，寻求经济、高效的答案。东京也许就需要这样直接的答案。“失格建筑”不是被文化熏陶的“建筑”，而是作为事物的“建筑物”。它们对于功能性的追求已经不是在比例上进行调整的问题。它们仅仅具备功能本身，好像把那种不在乎众人评判的态度原原本本地表达出来。它们毫无表现欲，平常得不能再平常。例如，把混凝土工厂和混凝土搅拌车司机的住所合并到一起，形成“工作生活一体”的终极形态；但从建筑规划的常识来看，工作区和居住区应当分离，这么做不过是恼人的功能拼凑。不过，东京本来就是一个充满悖论的城市，正是这些“建筑物”反映了东京城市空间的本质。试图用故事性的历史和文化来装点东京的做法反而让人觉得不真实。

“失格建筑”对文化不敏感，却十分务实。楼间空隙、屋顶、墙面、土木结构，无论什么都使用现场已有的材料。重要的是，这些建筑发现了赋予一个建筑要素双重功能的使用方法。例如，高速公路的下方、仓库屋顶的上方、建筑物和建筑物之间的空隙之类的空间副产品，改变使用方法，就能开发出新的功能。这一点也许可以被称作城市环境中的“可供性”。从结果看，像是“高速公路＋百货商店”这样建筑类别完全不同的复合体诞生了。百货商店这个结构体依托高速公路而建，而高速公路这

be opened up by the buildings made in Tokyo?

From <Architecture> towards <Building>

The buildings of Made in Tokyo are not beautiful. They are not perfect examples of architectural planning. They are not A-grade cultural building types, such as libraries and museums. They are B-grade building types, such as car parking, batting centres, or hybrid containers and include both architectural and civil engineering works. They are not 'pieces' designed by famous architects. What is nonetheless respectable about these buildings is that they don't have a speck of fat. What is important right now is constructed in a practical manner by the possible elements of that place. They don't respond to cultural context and history. Their highly economically efficient answers are guided by minimum effort. In Tokyo, such direct answers are expected. They are not imbued with the scent of culture; they are simply physical 'building.'
Moreover, Tokyo is really such a contradictory place, because it is in fact these buildings which most clearly reflect it's quality of urban space, whereas the translation of issues of place through history and design seems like a fabrication – This is Tokyo.

Where cultural interest is low, interest in practical issues is high. Whether civil engineering structures, rooftops, walls or gaps between buildings, utilise whatever is at hand. What is important is the discovery of how to establish a second role to each environmental element. With this doubling up, it becomes possible to re-use spatial by-products. The material is not given, but is discovered through our own proposition of how to use it. It might be termed 'affordance' of the urban environment. Further to this, cross categorical hybrids such as expressway-department stores can arise. In this example, the department store depends on the expressway for its structure. On the other hand, the expressway depends on the department store for its validity in such a busy commercial area. So neither can exist on their own – they are interdependent.
Such existence seems anti-aesthetic, anti-historic, anti-planning, anti-classification. It releases the architecture of over-definition towards generic 'building.' The buildings of Made in Tokyo are not necessarily after such ends, but they simply arrive at this position through their desperate response to the here and now. This is what is so refreshing about them.

个欠缺美感的结构体处于商业区的正中心。这个例子里的两个建筑物无法单独成立，谁也离不开对方。

这样的方法也许反美学、反历史、反规划、反分类，却解放了自我封闭的“建筑”，使之成为百无禁忌的“建筑物”。东京制造的建筑物并不刻意追求出格，只是拼命地想要解决“此时此地”的问题，于是就自然而然地长成了这副充满生机的模样。真是让人耳目一新。

邻接性和“环境单元”

无视建筑类别的复合体制造出了事物之间未曾预想的邻接性，无关的功能被放进同一个结构之中，邻接的若干建筑物与结构物被作为一个整体使用，一座建筑物就包含了和城市之间独特的生态关系……在这么多案例之中，我们关心的问题是如何在城市环境中制造各种各样的“统一体”，以及如何从这些“统一体”中找出使用方法，这就是城市的生态问题。

过于密集的东京催生了不受建筑物轮廓限制、挣脱建筑类别束缚的“统一体”。它们既没有固定的轮廓，也没有既定的单元或规模，无法被定义为建筑、土木工程、城市、景观中的任何一项。因为它们是由邻接的环境形成的一个统一体，所以我们把它们称作“环境单元”。它们在废弃和再造的过程中循环，无法保留类似临街外立面这样长期存在的稳定结构。邻接事物的表 / 里、内 / 外也在再造的过程中不断发生着转变。东京的建筑物就处在这样一种流动的状态之中。

建筑师普遍的价值观不允许他们造出如此多样的“统一体”。把建筑分门别类，规划物理结构，规定建筑功能，经过这样一番流程完成的建筑才符合他们的常识。这是现代主义留下的遗产，时至今日，这种方法已经变得无比精确。但是，日常生活告诉我们，建筑物不需要如此精确。生活空间无法单靠一座建筑物成立，而要靠若干建筑物及其邻接的周围环境共同构成。为什么不认可这种因为过度密集而产生的城市现象的价值，并畅想它们未来的模样呢？如此一来，公共建筑将不用建在破坏周边环境连续性的空地上，不用再被打包进一个单独的箱子，而是可以变得更小、更分散，形成一张把邻接的环境都容纳进来的网络。

Adjacency and 'Environmental Unit'

Our interest is in the diverse methods of making and using coherent environments within the city, together with the urban ecologies seen there. This includes the unexpected adjacency of function created by cross categorical hybrids, the co-existence of unrelated functions in a single structure, the joint utilisation of several differing and adjacent buildings and structures, or the packaging of an unusual urban ecology in a single building.
In Tokyo's urban density, there are examples of a coherency which cross over categorical or physical building boundaries. It is something which differs from the architecture of self-standing completeness. Rather, any particular building of this kind can perform several roles within multiple urban sets. They cannot be specifically classified as architecture, or as civil engineering, city or landscape. We decided to name such coherent environments of adjacency 'Environmental Units.'
Furthermore, the external envelope does not act to divide public and private, as in the traditionally understood idea of a facade. We are in a fluid situation, where rigid distinctions such as between shallowness and depth or front and back, are easily overturned by a shift in the setting of the ecological unit.
The magnificent Architecture of Architects retains distinctions between categories, rationalises physical structure, pushes preconceived use onto that structure, and tries to be self-contained. This is even though there are so many diverse ways to define environmental unities. It is a method that Modernism has passed down to us, and the precision of its ways is becoming stronger and stronger. Yet, everyday life is made up of traversing various buildings. Living space is constituted by connections between various adjacent environmental conditions, rather than by any single building. Can't we draw out the potential of this situation and project that into the future? If we can, it may be possible to counter the typical Japanese Modernist public facilities which are cut off from their surroundings and packaged into a single box. We can place attention on the issue of how usage (software) can set up a network, where public facilities can be dispersed into the city whilst interlapping with the adjacent environment. Spaces for living can penetrate into various urban situations and thereby set up new relations amongst them. The possibilities for urban dwelling expand.

ON/OFF（开 / 关）

“环境单元”的统一体是由三种不同的秩序共同创造出来的：按类别分类的秩序、物理结构的秩序，以及建筑功能的秩序。还是以“高速公路＋百货商店”的复合体为例，上方的交通和下方的购物共享同一个结构体，但它们的类别不同，功能上也没有任何关联。也就是说，这座建筑物只在结构这一秩序中才成为一个统一体。它之所以被我们视为“失格建筑”，既不是因为我们不知作何评价，也不是因为它没有名气，而是因为它包含的建筑类别不一致（建筑和土木工程）、建筑功能不统一（高速公路和百货商店），只勉强靠结构形成一个整体。这种岌岌可危的结合给人“失格”的感觉。我们尝试把遵循秩序视为 ON，违反秩序视为 OFF，那么东京制造的建筑在类别、结构、功能这三项秩序中至少有一项是 OFF（图 1）。与之相对，这三项秩序都是 ON 的建筑，多是由大型设计事务所或承包商担当设计的，大都有完整而清晰的规划。再者，这个图表里唯一含有 OFF 但又不符合东京制造标准的部分（类别 ON、结构 OFF、功能 OFF），对应的例子是巴黎街景。通过东京制造的建筑试图去寻找的，正是整洁的现代主义建筑和巴黎街景之间存在的可能性。现代主义城市空间和巴黎街景常常被看作两个极端，但它们只是这里提出的 ON/OFF 可能性的一部分而已。

此外，ON/OFF 的问题还涉及材料、设备等方法的等级的高 / 低、建筑类型的 A 级 / B 级、审美判断上的美 / 丑、历史价值的有 / 无、规划道德方面的善 / 恶。包含这多重判断标准的 ON/OFF 组合的“乐章”，真可以用“爱恨交织”来形容，它们反映了时代心理的起伏，表现了建筑物微妙的社会等级。如果把所有的判断标准都定为 ON，那建筑物就只有一种形态了。但如果允许出现 OFF，建筑物的可能形态就会呈指数增长。仅取类别、结构、功能这三个标准，如果允许 OFF 存在，可能的形

On/Off

We can find an overlapping of 3 orders which set up the 'Environmental Unit.' They are based on category, structure and use. If we take again the example of the hybrid between expressway and department store, the traffic above and the shopping below are simply sharing the same structure, but belong to different categories and have no use relation. In other words, it is only structural order which unites this example. Maybe it is not that the example is impossible to evaluate within the existing cultural value system, or the norm for architecture. Rather, the sense of unity is full of dubiousness which is the essential reason that this example is da-me architecture. We can say that when any of the 3 orders are operating, they are 'on,' whereas when they do not take effect they are 'off' (fig. 1: Made in Tokyo Chart). This system starts to incorporate all the value poles which seem to form such an important role in the recognition and indeed the very existence of da-me architecture. We can recognise that the examples of Made in Tokyo almost always comprise some aspect of being 'off.' The only vacant endpoint to the Chart that includes an aspect of 'off,' is the position which might be filled by the continuous street facades of Paris. By contrast, the magnificent buildings of Architects are 'on,' 'on,' 'on.' Often, the Parisian streetscape and the Modern city are held to be in opposition, but the abundant examples of Made in Tokyo show that they are not necessarily bipolar. They simply exist within a score of on and off.

Anyway, surely too much 'on' can't be good for our mental landscape. If we switch all 3 orders 'on,' there is only one possibility for achieving satisfying architecture, but if we allow any or all aspects to be 'off,' then suddenly the possibilities for variation explode to 8 (2 to the power of 3). This establishes a huge release for those who are designers. When we say that we can sense the pulse of Tokyo in the 'da-me architecture' which includes some aspect of being 'off,' it means that even though the urban space of this city appears to be chaotic, in exchange, it contains a quality of freedom for production. The landscape of Tokyo is a random layering of different buildings corresponding with multiple social purposes. We hope in our design work to clearly represent possibilities for the urban future by being consistent with the principle findings of our research. The observations can only gain a certain clarity once they have been studied through design and vice versa. Such interactive feedback between observation and design is one efficient method through which to contribute to the city through the scale of architecture (fig. 2: On/Off table).

态就有 8（2 的 3 次方）种。这对于设计师来说是莫大的自由。东京的城市空间表面上十分混乱，但这意味着其内部允许这样的自由。对应着各种社会性的建筑物错乱分散地重叠在一起，形成了东京的风景。因此，我们确实从含有 OFF 元素（被认为是理解社会性构造的抓手）的“失格建筑”里，感受到了东京的可能性。再进一步，如果能把这些观察运用到实际的设计中，在当下的建筑物中清晰地展现这座城市的未来形态，那就再好不过了。对城市的观察首先需要借助建筑设计来明确态度。如果我们希望通过建筑为城市做贡献，那么这样的观察和设计相互依存的建筑实践是颇为有效的办法（图 2）。

从城市内部出发

这本导览手册无意将东京的一切一网打尽，恐怕也难逃十年之后过时的命运。但我们也发现，把东京这样一座巨大的城市作为研究对象，本身就是一个不可能的任务。不过，即使在这样的城市里，仍旧可以找到以建筑物为中心的环境统一体。我们可以从城市内部的建筑出发，观察生活于其中的人们——建筑的所有者、使用者和来来往往的人。这种观察反映了我们对今日东京的身体与心理感受。当我们以使用者或设计师的身份走在城市中时，它可以帮助我们形成对城市的个性化理解。我想，只有从城市内部出发，我们的建筑冒险才能一路向前。

图 1：东京制造图表　fig.1: Made in Tokyo Chart

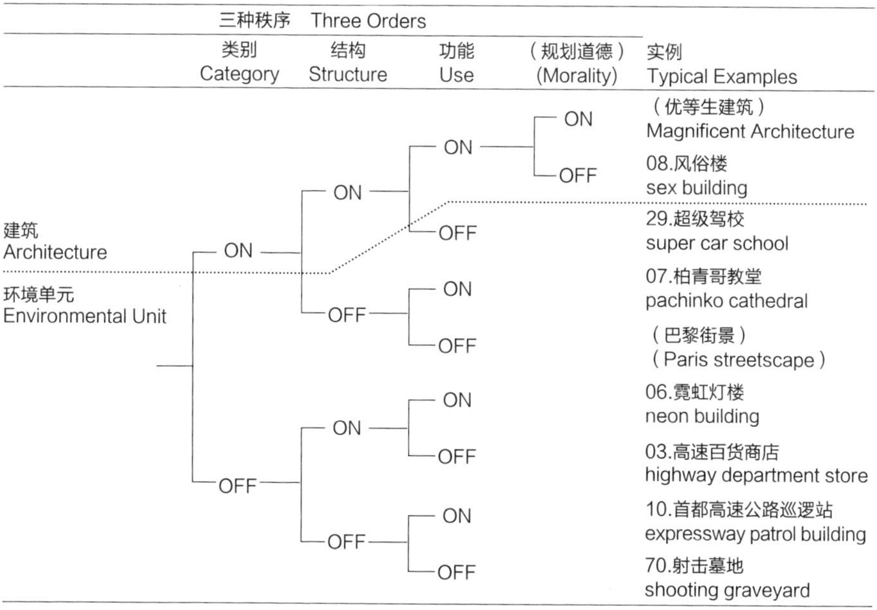

	三种秩序 Three Orders				
	类别 Category	结构 Structure	功能 Use	（规划道德）(Morality)	实例 Typical Examples
建筑 Architecture	ON	ON	ON	ON	（优等生建筑）Magnificent Architecture
	ON	ON	ON	OFF	08.风俗楼 sex building
环境单元 Environmental Unit	ON	ON	OFF		29.超级驾校 super car school
	ON	OFF	ON		07.柏青哥教堂 pachinko cathedral
	ON	OFF	OFF		（巴黎街景）(Paris streetscape)
	OFF	ON	ON		06.霓虹灯楼 neon building
	OFF	ON	OFF		03.高速百货商店 highway department store
	OFF	OFF	ON		10.首都高速公路巡逻站 expressway patrol building
	OFF	OFF	OFF		70.射击墓地 shooting graveyard

From Inside the City

This guidebook which captures the living condition of Tokyo, may seem to be old in 10 years time. But it is impossible to attempt to take on the whole of the megalopolis of Tokyo. Yet from the scale of a building, inside the city, it must be possible to see owners, users and passers-by. It is possible to find environmental units with buildings at the centre, within this never ending city. This can become a bodily grasping of our understanding of urban reality. We think that our architectural adventure can only start from here.

No.	名称	name	类别 Category	结构 Structure	功能 Use
1	仓库球场	warehouse court	○	○	●
2	电器走廊	electric passage	●	○	●
3	高速百货商店	highway department store	●	○	●
4	电影桥	cine-bridge	●	○	●
5	过山车楼	roller coaster building	●	●	●
6	霓虹灯楼	neon building	●	○	○
7	柏青哥教堂	pachinko cathedral	○	●	○（发生的关联）*1
8	风俗楼	sex building	○	○	○（单独增殖）*2
9	卡拉OK宾馆	karaoke hotel	○	○	○（单独增殖）
10	首都高速公路巡逻站	expressway patrol building	●	●	○
11	大使馆楼	embassies building	○	○	○（单独增殖）
12	公园停车场	park on park	●	○	●
13	公交住宅区	bus housing	○	○	●
14	高尔夫出租车站	golf taxi building	●	○	●
15	混凝土公寓	nama-con apartment house	●	○	●
16	汽车塔	car tower	●	○	○
17	马公寓	horse apartment house	○	○	○（发生的关联）
18	物流综合体	distribution complex	●	○	○
19	空调楼	air-con building	○	○	○
20	广告公寓	billboard apartment house	●	○	●
21	神社楼	shrine building	○	●	●
22	渣土公寓	sand apartment house	○	○	●
23	配送螺旋	delivery spiral	●	○	○
24	澡堂旅游楼	bath tour building	○	○	○（发生的关联）
25	出租车楼	taxi building	○	○	○
26	货车塔	truck tower	○	○	○
27	立交球场	interchange court	●	○	●
28	跃层式加油站	double layer petrol station	○	○	●
29	超级驾校	super car school	○	○	●
30	污水球场	sewage courts	●	○	●
31	净水球场	supply water courts	●	○	●
32	墓地通道	graveyard tunnel	●	○	●
33	阿美横町空中寺院	ameyoko flying temple	○	●	○（发生的关联）
34	商店崖	shopping wall/mall	●	○	●

37	增殖滑道楼	proliferating water slides	○	○	○（单独增殖）
38	通风方尖碑	ventilator obelisk	●	○	○
39	车站上的家	apartment station	●	○	●
40	蜈蚣住宅	centipede housing	●	●	●
41	汽车村	vehicular village	●	●	○
42	潜水塔	diving tower	●	●	○
43	底盘公寓	chassis apartments	○	○	○（单独增殖）
44	TTT（乐高办公楼）	TTT (lego office)	○	○	○（单独增殖）
45	隧道神社	tunnel shrine	●	●	●
46	公寓山寺	apartment mountain temple	○	●	●
47	吸血公园	vampire park	●	●	○（发生的关联）
48	起重机架	crane shelves	○	○	○（单独增殖）
49	幽灵铁道工厂	ghost rail factory	●	○	●
50	挡土墙公寓	retaining wall apartments	●	○	●
51	桥洞里的家	bridge home	●	○	●
52	宅地农场	residential farm	●	●	○
53	物流立交枢纽	dispersal terminal	●	○	○
54	皇家高尔夫公寓	royal golf apartments	●	○	●
55	汽车办公楼	car parking office	○	○	○
56	绿色停车场	green parking	●	●	○
57	汽车百货商店	auto department store	○	○	●
58	家庭餐厅三兄弟	family restaurant triplets	○	○	○（发生的关联）
59	果蔬小镇	vegetable town	●	○	○
60	松茸形车站楼	sprouting building	●	○	●
61	TRC（东京物流中心）	tokyo dispersal centre	○	○	○
62	冷冻园区建筑群	coolroom estate	○	●	○
63	宠物建筑1号	pet architecture 001	●	●	●
64	水坝公寓	dam housing	●	○	●
65	立交枢纽航站楼	airport junction	●	○	●
66	运动桥	sports bridge	●	○	●
67	运动者动物园	sportsman zoo	○	●	○
68	直升机仓库	heli-warehouse	○	○	●
69	洗车露台	carwash terrace	○	○	○
70	射击墓地	shooting graveyard	●	●	●

图2：ON/OFF表格　fig.2: On/Off table

图例　○=ON　*1 generative relationship　*2 monoproliferating　●=OFF

导览手册的形式

我们的搜集所得既是东京这座城市的调查报告，也是一份了解东京城市空间的导览手册。尤其对于来自国外的旅行者而言，这是一册颇为独特的东京旅行指南，所以我们特意制作了日英双语版本。建筑物的序号基本上按照我们搜集的顺序排列，没有其他特殊的含义。这是一本无始无终的导览手册，通过这种形式我们把东京呈现为一个平面。

调查方法

我们先凭着记忆把平常留意到的建筑物做成一张大致的清单，同时使用各种交通手段亲身造访这些建筑物。以此为基础，我们开始讨论何为“东京制造”，然后把更多的建筑物加进清单。从若干大家都认可的建筑物之间找到共通点后，评选的标准也就渐渐固定下来，这一过程好像在城市里自由策展。从结果看，我们寻找的都是那些带有 OFF 感的建筑物，范围不局限于东京都内，也延伸到了东京近郊。

METHOD
东京制造的方法

Guidebook Format

This collection is a project researching the city of Tokyo. At the same time, it is a guidebook to understand an aspect of Tokyo urban space. In particular, it can become a unique navigation tool for foreign visitors , and so we have endeavoured to make it bi-lingual. The buildings gathered here are not ordered by any storyline. By using the endless format of the guidebook, we thought to present Tokyo as an expanding field.

Research Method

We made an initial list of buildings which had stood out in our everyday life. We flicked through the city on the back of trunk routes, as well as various other modes of transport such as rail, ferry, bicycle. And we discussed the question of 'what is Made in Tokyo?' The list became thicker. The process of fixing the selection criteria involved connecting convincing examples and discovering their commonalities. It is a kind of urban curation. As a result, we realised that what we were looking for was buildings with a sense of 'offness.' The extent of our study expanded into the greater urban area of Tokyo. Anyway, by constantly moving, we could gain a slight distance from the burying containment of the everyday city. Observation was not focussed on the building itself, but was with a slight 'zoom back.' We tried to view the full panorama – the building and the surrounding environment together – to see another facility. For the moment, we forgot the categorical divisions between architecture, civil engineering, geography, and sought to see things as simple, physical unities. Flows of people, infrastructure is the object of our observation. It is possible, amongst that flow, to find small eddies. Usually, built structures must

我们最常用的交通方式是自驾。以首都高速公路、环状 7 号线、环状 8 号线、国道 16 号、国道 17 号等主干道为中心，特别留心周围的准工业区、邻近商业区等功能高度混杂的地域。自驾之外，我们还利用火车、轮船、自行车和徒步等方式，尝试以各种不同的速度、高度、角度穿过东京。有时我们也会爬上高层建筑进行鸟瞰式的观察。为了搜集目标而徘徊于城市之中，调动起全身的感官，不放过任何微小的迹象，这让我想起在森林中捕捉独角仙和锹甲虫时的情景。尽量迅速地连续移动，就可以与日常拉开一点距离，眼前就会出现新的风景。

观察的时候并非直接注视建筑物本身，而要后退一步，关注建筑物的全景。如果可以的话，把建筑物周边的环境也包括进来，再把这个整体作为一个新的设施考虑。在城市里一边移动一边纵览时获得的距离感，和这样后退一步，把视角拉远的做法如出一辙。

不要用力地盯着建筑物，而是要不经意地去看。先忘掉建筑、土木工程、构筑物、地形这些概念，只是把建筑当成一个物理上的统一体。无论哪座建筑物，都是参与构成城市空间的物理结构物。

如果把城市环境看作人群和基础设施的流动，统一体就如同这些流动中旋涡一般的存在。人工的结构物大体上都是技术上的完成品；但是，建筑物的使用与居住都不受技术上的完成度束缚。人的活动可以横跨多个建筑物完成，反过来，人的活动也受到建筑物和土木构筑物等物理结构的限制。按常理应该被分隔到不同空间的活动在同一个空间里进行，这样的情况也很常见。意想不到的生态关联和相互依存的关系就在这样的空间里发生。通过这种方法观察城市，就像绘图一般从东京的土地上提取出一个个有意义的环境统一体。

制作材料

○照片：拍摄照片是为了发现建筑并当场记录下最初的观察情况。

○绘图：但仅靠照片无法把观察到的东西都记录下来，还需要绘制各个建筑的图解。用单线绘制等角立体图，再拉出线说明构成要素、内部结构以及周边环境。在整理说明的过程中，对每座建筑本身的信息做了取舍，这样观察的标准就不会模糊不清。细致地绘图，有助于我们怀着爱去观察建筑。

○地图：用于显示建筑物的规模、形状以及与周边环境的关系。当然，地图也是导览手

hold a certain technological completion. But to use it, or live in it, is not necessarily related to such completeness. Any single activity can stand astride several structures at once. But on the other hand, our activities are regulated by physical structures such as architecture and civil engineering. Several inexplicable activity adjacencies can also be seen in the same building. Through such a particular process of looking, our eyes can be trained to pick out certain figures from the ground of Tokyo. A summary of our approach might be to say that we 'zoom back' looking for 'cross-categories' and 'urban ecologies.'

Material Data

Photos: We took photos to make a quick and immediate record of our initial thoughts and discoveries.
Drawings: But photos cannot contain all of our observations. So we made drawings to figure out the actuality of each example. Each was drawn in single line isometric, and the elements and internal structure and related environment are explained by notes. We tried to limit blur in observational criteria by organising the format of explanation. Careful drawing helped us to see the object of our study with love.
Maps: In order to show the scale and shape in contrast with the surroundings, we inserted the site plans in maps. Of course, they also function as guidebook maps.
Nicknames: We gave nicknames to each discovered example, to immediately explain where the interest in the building is, and to express our fondness. It is also a signature, training these buildings with no author to become pieces of architecture to the urban curator.
Text: Each example has a number. Basically, the buildings with earlier numbers are the ones discovered earlier. Of course, because it is a guidebook, we also need addresses. We recorded the functions included in each, to explain how they

册的一部分。

○昵称：我们满怀着喜爱之情为这些建筑取了昵称，能让大家迅速领会到它们的趣味。昵称也是一种签名，是观察者让这些也许作者不明的建筑成为作品的方式。

○文字信息：为了整理搜集到的材料，使其成为研究资料，我们给每座建筑物都加上了编号。一般来说，编号越小，发现得越早。当然，因为是导览手册，所以还附上了地址，这能让读者迅速了解街区的氛围。为了说明建筑物的用途，我们列举了每座建筑的功能，于是发现了许多令人意外的功能组合，许多新的建筑类型。为了让读者更好地了解这些城市里的“插曲”，我们又加上了现场观察的细节说明。

○“东京制造”地图：把这 70 座“东京制造”的建筑排列在一个平面上，就构成了一幅虚拟的东京浮世绘。

媒介

把这些爱恨交织的观察用某种媒介固定下来，观察就变成了一股力量。“东京制造”的发现一边在不同媒介上持续更新，一边催生了形态各异的产品。以散步时随手拍摄的照片（1991）为起点，在东京的年度建筑展（1996）上第一次展示了“东京制造”的调查过程。当时，我们把搜集到的 30 座建筑做成了目录和 T 恤。在 T 恤正面印上建筑照片，在肩膀的位置印上建筑图解，将价格标签做成导览手册，还设计了一个“东京制造”的织唛。建筑物通过 T 恤被穿在身上，随之重新回到城市，不断增殖。这个展览后来在苏黎世举办（1997），我们制作了瑞士版的 T 恤。搜集到的建筑增加到 50 座；在大日本印刷和传媒设计研究所的协助下，又上线了日文网站（1998）。这一时期，我们还制作了描绘城市微观生态系统的动画短片。被邀请参加威尼斯国际建筑双年展的“Expo on Line”单元时，我们上线了英文网站（1999）。这次，在鹿岛出版会的协助下，《东京制造》的纸质书得以出版。

随着内容增加、媒介变换，我们锻炼了观察城市环境的眼力。这些探索既适用于观察东京这座城市，也为思考如何观察城市并付诸实践提供了思路。我们将沿用导览手册的形式，让此刻的城市调查通向未来。

are being used. These records show unforeseen building types with unexpected combinations of function. The detailed comments help to clarify how the example works within a series of urban episodes. Made in Tokyo Map: A virtual ukiyoe map of Tokyo composed of all 70 Made in Tokyo examples.

Medium

Fixing observation in conjunction with both love and hate emotions is in effect changing observation into a power. The survey work of Made in Tokyo has given rise to several differing products each time it appears in differing media. First, it was snap photos while going for walks (1991), and then it took shape as a survey as part of the exhibition 'Architecture of the Year 1996', where it appeared as 30 examples expressed through a catalogue and t-shirts. The t-shirts had photos on the front, drawings on the shoulder, take away guidebook sheet as a price tag, and we made an original Made in Tokyo brand tag. We aimed to multiply the effect of the buildings by throwing them back into the city on the bodies of people. The exhibition also travelled to Zurich, and we made a Swiss version of the t-shirts. The examples increased to 50, in making the Japanese version of a home page (1998). At this time, we animated micro-urban-ecologies. When we were invited to take part in 'Expo on Line' of the Venice Biennale, we prepared the English version of the home page (1999). Now Made in Tokyo has become a book.

Increasing the contents and changing the medium builds up a way of looking at the urban environment. Such work in progress seems very suitable for studying the city of Tokyo. The guidebook format allows the possibility of this kind of lack of finality. The survey will continue. Made in Tokyo is still swelling to this very day.

10 KEY WORDS
东京制造的十个关键词

跨界格斗
CROSS-CATEGORY

自动尺度
AUTOMATIC SCALING

宠物尺寸
PET SIZE

物流城市
LOGISTICAL URBANITY

运动场
SPORTIVE

副产品
BY-PRODUCT

城市居住
URBAN DWELLING

作为建筑物的机器
MACHINE AS BUILDING

城市生态系统
URBAN ECOLOGY

虚拟用地
VIRTUAL SITE

跨界格斗

1976 年 6 月 26 日，安东尼奥 · 猪木和穆罕默德 · 阿里进行了一场跨界格斗比赛，目的是验证摔跤和拳击哪个更强。比赛以平局收场，而且还出现了猪木躺在垫子上踢阿里的腿这样无聊的画面。不过，正是这样的跨界格斗打破了以往封闭自身以求和其他竞技严格区分发展方向的近代体育趋势，让摔跤运动回归“格斗”的原点。

现代城市的形成也伴随着劳动分工和专业化的过程。最明显的例子要数建筑与土木工程，或建筑规划与城市规划的区分，政府机关和大学也沿用了这样的区分。这引发了许多问题，其中最大的问题就是，大家无法把建筑作为一个整体来考量了。同一座建筑被形形色色的立场割裂，参与者各说各话，因此导致环境恶化、资金浪费，这样的例子不在少数。大家应当跨过建筑、土木工程、城市规划，甚至造园、农业、林业这样既有分类的限制，来建设与维护自己日常生活的环境，这就是城市建设的跨界格斗。如果能够摒弃类别的限制，为构造一个统一的环境准备的工具就能大幅增多。

东京制造的建筑物多是这种跨界格斗式的复合凝缩。“2. 电器走廊”“3. 高速百货商店”都是由高架铁路或高速公路和下方的商店街或百货商店组成的大型建筑物。交通结构物和建筑杂合，共同形成一个统一的结构物，但这些细长的建筑物本身没有明显的轮廓，而是不断延长，直到消失在视野之中。对于同一个物理结构物来说，因为有铁路和公路，可以说它是交通结构物；因为有百货商店，又可以说它是建筑，难以对其下一个单一的定义。在下方的街道散步时，人们意识不到头顶上有车驶过；反过来，上方车里的人也不会对下方的商店有多少了解。这样的建筑组合按常理是不应该存在的，只不过铁路和公路的细长物理形状刚好匹配商店街的需求罢了。它们功能不同，用途各异，仅因毗邻商业区这一外部条件，就决定了这块细长区域的使用方法。结果，这一类高架下的复合建筑物不再被单一的服务目的占据，多数情况下都能将回应周边环境的需求也纳入其中。

“29. 超级驾校”在超市的屋顶铺设柏油路面、安装路灯，开起了驾校。在一般的常识里，汽车应该在路上行驶，于是就把建筑物的屋顶修成公路的样子。这与意大利的理性主义建筑物“林格托”颇为相似。林格托是菲亚特的汽车工厂，楼顶修建了试驾跑道。但在超级驾校

CROSS-CATEGORY

26 June in 1976, there was a highly publicised fighting match between Muhammad Ali and Antonio Inoki took place. I was very keen to discover whether boxing or pro-wrestling would be stronger, but the result was a draw, and the actual fight was not even interesting as Inoki just made low kicks from the mat to Ali's legs. However, this kind of cross-category match breaks through the self-referential structure of Modern fighting sports. Instead of each type becoming more and more clearly defined through comparison with each of the others, it tries to return to the essence of fighting.

Modern city planning is of course not the same as fighting sports, but it also evidences separations and specialisations. The strongest separations are between architecture and civil engineering, or between architectural and urban planning, and these can be seen throughout bureaucracies and the academy. There are many problems stemming from this situation, but the most immediate is the fact that a single location cannot be thought through in its totality. There are countless instances of the environment in fact being aggravated by being fed with uncoordinated ideas from differing fields, let alone monetary wastage. So it would be good if we could create and maintain our own environment by losing the strict definition of such categories as architecture, civil engineering, urban planning, as well as advertising, agriculture and geography. This is the cross-category match of urban production. If the categories can be cross-breed, the tools for organising a coordinated environment can suddenly increase manifold. This kind of cross-categorical hybridisation is already condensed into the buildings of Made in Tokyo.

的例子里，建筑物的上下两部分没有功能上的联系。只不过因为驾校的存在，车站附近一直很热闹，超市抓住机会开在了这里。芝浦的“68. 直升机仓库”利用了仓库靠近海岸、视野开阔的优势，在其上修建了直升机停机坪，而由直升机运输的物资并不交给下方的仓库保管。在这种交通设施和建筑杂合、彼此毫无关联的多层建筑物里，各项功能分别和周边的环境联结，各有其存在的理由。但它们之所以会被收纳到同一个结构物之中，通常只是因为所需建筑平面的大小刚好差不多罢了。

在构筑物和建筑的杂合里也可以发现类似的例子。位于羽田的“20. 广告公寓”在十层的公寓楼上架起了超市的巨型广告牌。这块广告牌足有五层楼高，相当于整体高度的三分之一，视觉上明显感觉比实际更大，好像公寓是广告牌底座之类的附属物似的。但如果有人循着这块广告牌找过来，就会发现此处根本没有超市。这为什么没有成为问题呢？因为这块广告牌正对着羽田机场的跑道，让飞机里的乘客从飞机舷窗眺望时能够看到才是架设这块广告牌的目的。这类构筑物和建筑组成的杂合体，通常都是由互不相关但大小差不多的元素拼接而成的，看起来就像是生拼硬凑出来的一样。

互不相关的多层建筑甚至也出现在神社、寺院等宗教场所里。“32. 墓地通道”就把墓地底下挖空修建了隧道，有传闻说此地经常有幽灵出没，因此也被称作“幽灵隧道”。“45. 隧道神社”因为东北新干线开通的缘故，变成了位于隧道上方的神社。这是后来规划的交通网和旧道路结构之间的冲突，隧道神社把这样的矛盾原原本本地保存了下来。只从结果看，这两个例子都是交通结构物和墓地、神社或寺院组成的互不相关的多层建筑。在“34. 商店崖”的例子里，山崖的岩壁被改造成建筑的一部分，可以把它看成建筑和地形的杂合，或是建筑和挡土墙的杂合。

我们无法把构成环境统一体的建筑、土木工程与地形的杂合体放到某个具体的类别中，也无法将杂合体拆解后把它们看成单纯的景观。所以，我们姑且把这样的统一体称作“环境单元”。它们不存在某个类别内部的技术体系意义上的秩序感，而是利用长度、面积等物理上的尺度，以及和邻接环境之间几何形态上的对应关系，创造出有意义的统一环境。

自动尺度

也许是因为人口和资源的集中抬高了东京

Examples of cross between infrastructure and architecture:
2. electric passage
3. highway department store
29. super car school
68. heli-warehouse
Examples of cross between commercial structure and architecture:
5 . roller coaster building
6 . neon building
20. billboard apartment house
Examples of cross between constructed ground and architecture:
32. graveyard tunnel
45. tunnel shrine
Examples of cross between retaining wall and architecture:
34. shopping wall/mall
50. retaining wall apartments
64. dam housing
Even in the landscape of Tokyo. which is so often claimed to be 'chaotic', a certain environmental coordination made up of categorical crosses between architecture, civil engineering and geography, can be found. There is a clear logic in the way that differing activities are brought together by physical convenience such as scale and adjacency. We can see that part of Tokyo's dynamism is ordered through physical terms rather than the categorisation of contents. We start to recognise the unexpected interdependence of activities by looking at Tokyo in this kind of positive way – as a cross-category match of urban production.

AUTOMATIC SCALING

Because of inflationary land prices, there is a 'void

的地价，在东京看到空地，就会让人产生“真浪费啊”的念头。“空白恐惧症”在东京蔓延，因此无论走到哪儿，都可以看到填满这些空隙的尝试。塞满空隙的这些东西没有功能上的关联，它们被聚集在一起只是因为它们与空隙大小相同、面积相当。我们把功能各异的东西因为规模而自动聚集到一起的现象称为“自动尺度”。它们大胆超越城市的历史和社会形式的既定规则，将矛盾却实用的建筑拼接在一起。这样一根筋地把空白填满的想象力、智慧和努力，更新了建筑结合的方式。

“2. 电器走廊”是总武线秋叶原至神田之间的高架铁路下塞进了一排小型电器商店形成的空间。每间店铺的纵深由高架的宽度决定，根据大小的不同，有的店跨一个桥拱，有的店跨四个桥拱，填充了高架桥墩之间的空隙。如此，小型电器商店切分了桥墩之间的长条空白，在高架这一铁路交通的副产品下形成一条不逊于百货商店的电器走廊。“5. 过山车楼”是一座沿白山大道而建，宽 20 米、长 150 米的建筑物，它是后乐园游乐场的大门，利用长条状的屋顶搭起了过山车的轨道。到底是因为要建过山车的轨道才有了底下这座开有纪念品商店和餐厅的长条建筑物，还是先有了这座建筑物再在上面架起了轨道？谁也说不清，而这正是它的有趣之处。这里的过山车似乎在和白山大道上奔驰的车辆竞速，真是妙趣横生。

“13. 公交住宅区”是在都营公交的车库上建起的板状的都营住宅。一个结构单位能容纳两辆公交车，上方的居住空间则容纳了两户住户。也就是说，住宅的长宽几乎和公交车一样。如果都营住宅的标准尺寸真的是参考公交车的尺寸制定的，那多少有些黑心啊。

在“跨界格斗”一节里提到过的“29. 超级驾校”是在超市的屋顶开起了驾校，这两者毫不相关，只是偶然占用了同一块空间，这是典型的超理性主义建筑。

“39. 车站上的家”是在私营铁路流山线车站的月台上建起的集合住宅，一层有出租车车库，月台与住宅楼长度相同，上方就是集合住宅的公用走廊。车站月台和住宅走廊恰好都是呈细长形状的露天空间，所以被放到了建筑的同一位置。家和车站零距离，虽然多少有点吵，但毕竟站在自家的走廊就可以边聊天边等电车。“60. 松茸形车站楼”是一组低层轻型建筑，建在因铁路地下化改造而产生的地上长条空地上，就像从松树的根部长出的松茸一样。它充分利用铁路的宽度，发挥车站附近一带的地理优势，以线状陈列出超市、录像带租赁店、租车店等商铺。这在

phobia' in Tokyo, which instils a reaction of 'what a waste!' when we see any unused space. Everywhere, the desire to find and fill gaps can be seen. What occurs in these openings is not usually related to the function of the host facility, but rather answers to a super-rationalism where the filling is matched to the gap simply according to size and proportion. Let's call the idea of the chance meeting of differing objects, purely given by measurements, 'automatic scaling.' At this point is born the kind of building with unexpected practicality of adjacency and connection, boldly ignoring and jumping beyond the history and social mode of the city. The knowledge, invention and imagination summoned in fully utilising these spaces to the extent of stubborn honesty, makes possible new urban relations.

Examples of auto-width:
2. electric passage
5. roller coaster building
13. bus housing
60. sprouting building
Examples of auto-breadth:
20. billboard apartment house
29. super car school
Examples of auto-length:
39. apartment station

In other words, there is a prescription for 'void phobia.' We just need to find anything of the right size, and try to till the available space. The insight of the urban observer is tested in their sense and

私营铁路地下化改造的区域十分常见，快餐连锁店往往会开在这些地方。

就算在东京发现空白也不用担心，在街上寻找和空白差不多大的东西，不假思索地把它填进去就行了。如何寸土必争地填补空白，是摆在每位城市生活者面前的课题。

宠物尺寸

东京大街上的东西虽然大小各异，但和其他城市比起来，其最明显的特征是很多东西都是自动贩卖机尺寸的。自动贩卖机在东京随处可见，而在国外却几乎看不到。这既和城市的治安环境有关，也受到城市本身的空间感的影响。例如，在日本，人们不能在自家和邻地的分界线上动工，但如果空地狭小，当然要尽可能把房子往大了盖，于是建筑物与建筑物之间就产生了无人使用（但猫会使用）的空隙。因为东京高昂的地价，人们当然不想浪费这样的空隙，于是自动贩卖机就成了救世主。现在东京常见的形态各异（比如极薄或极矮）的自动贩卖机，正是为了填充各种各样的空隙而被设计出来的。除了自动贩卖机，人们还设计出了能填满空隙的卡拉 OK 包厢、自动停车场、广告牌等设施。把它们称为建筑，尺寸太小；称为家具，尺寸又太大。它们犹如城市里的宠物一般在建筑物内部和城镇角落自由出没，在所有地方都是同样的状态（超级室内软装?）。“东京制造”发掘了深潜在城市之中，发挥出微小尺寸，即宠物尺寸的效果的那些建筑。

“8. 风俗楼”是新宿歌舞伎町的一栋多租户的大楼，每层都开着不同类型的风俗店，同一层又被分出多个小房间。这些小房间无论怎么花心思，内部也只能放下一张床、一把椅子和一部电话。无论什么样的建筑物，如果被切分成一个个小房间，最后都会产生一种稳定状态：房间和房间令人目眩地连成一片。大到写字楼，小到铅笔楼，都可以被切分成一个个小房间，这是这类建筑的优势。因此，风俗店能开在大小、形状各异的建筑物里。“9. 卡拉 OK 宾馆”也基于同样的原理。卡拉 OK 包厢的大小只要能容纳电视显示器、K 歌设备、桌子和沙发就可以了。用作卡拉 OK 包厢的小房间的排列方式还参考了不少既有建筑的形式，会根据不同的情况变换排列组合的方式，例如：所有房间纵向分布的塔式，在二层外侧走廊横向分布的公寓式，甚至还有在高层内部走廊连通的酒店式。“11. 大使馆楼”的存在，是因为东京地价高昂，财力不足的国家的大使馆便需要共用一座建筑物。整栋楼如同享有治外法

efficiency in working out how to gain maximum utility from these void spaces.

PET SIZE

The city of Tokyo displays a whole range of sizes, but as a specific characteristic, items around the size of a vending machine can be pointed out. In Japan, vending machines are so abundant, but they are not nearly so visible in other countries. This difference may be due to the extent of public security in various cities, but it is also due to the nature of the sense of scale which exists in each location. For example, the urban code of Tokyo stipulates that neighbours must accept any new building work if it retains 500mm distance from the boundary. Of course everyone's site is too small, and so they try to build to the maximum possible extent. Tiny slivers of space between buildings, which can only be utilised by cats, are the result. With the high price of land in Tokyo, eventually these spaces become desirable for use. Maybe the vending machine is a kind of saviour in this situation. In particular, the very thin or very low proportions which can sometimes be seen are probably because of this tendency towards filling all gaps. Other types such as the karaoke box, car parking machines and signboards have also developed a unique size so as to be able to slip into these spare spaces. These items are a bit too small to be recognised as architecture, but a bit bigger than furniture. They are the kind of size which

权的联合国，墙壁和地板就相当于国界。想与邻国交涉，只要敲敲隔壁房间的门；厕所里的闲谈大概也算是一种国际交流（？）。这样没有隔阂、没有拘束的大使馆不正是全球化时代需要的建筑形式吗？以上都是建筑物内部被细分为小房间的例子。

“44.TTT（乐高办公楼）”是集装箱仓库的办公楼，几个集装箱令人意外地装上了钢架楼梯、铝窗、天线、太阳能板这些明显只属于办公楼的设施。集装箱什么都能装，所以如果装上窗户，通了水电，就变成了容纳人类的微型居住空间。“52. 宅地农场”建在练马区住宅用地正中的一块被遗留的农田里，是一个蔬菜摊。摊子虽小，却是连接周边居民和农田的重要枢纽。通常，收获的蔬菜会被运往市场，分配到市场里的商店，再销往各家各户；但这个蔬菜摊省去了所有中间的物流环节，直接把田里的蔬菜送到各家各户的餐桌上。正因为这片近郊的农田被遗忘了，这样的产地直销系统才能出现。微型蔬菜摊成了住宅用地和农田共存的关键，为这一地区赋予了独特的紧密感。“63. 宠物建筑 1 号”建在三条路交会处的狭小三角形空地上，外墙上靠着各种类型的自动贩卖机。正因为建筑物本身规模很小，自动贩卖机的存在感很强，犹如房子的武装一般。以上是外表看上去就是微型建筑的例子。和普通尺寸的建筑物相比，它们会让人产生独特的留恋之情、身体感觉和距离感。这些微型建筑之所以存在，多是因为公路、铁路或河川的偏移产生了狭小的空隙。作为城市与活生生的身体之间的接合点，宠物建筑代表的是一种非常微妙的内在表达，形成了一种独特的建筑类型。

微型 = 宠物尺寸，正因为小所以才能在城市里自由存在。填充东京这座城市的宠物建筑作为城市和身体之间的接合点，应该被用来个性化定制城市环境，使其更加宜居。

物流城市

如果想让国外的旅客看到东京原原本本的样子，就带他们上首都高速公路转一转。为配合 1964 年东京奥运会而紧急修建的首都高速公路在规划时以公共用地、公园、河

can exist in the corner of a room, or in the corner of the city, turning the urban environment into a 'superinterior'. The items' constant suppleness to fit their surroundings makes them like the pets of the city. So we can say that smallness = pet size.
Examples of spatial dice and mix:
8. sex building
9. karaoke hotel
11. embassies building
44. TTI (lego office)
52. residential farm
63. pet architecture 001
Smallness is a size which allows a freedom in urban action. If we consciously consider the abounding pet sized objects of Tokyo as an interface between the city and the human body, then our urban environment can become more and more comfortable.

LOGISTICAL URBANITY

If we wanted to show a visitor from another country the real Tokyo, then driving along the expressways would be highly recommended. Because the expressways were constructed in a hurry to be in time for the Tokyo Olympics, they are mainly sited over public land, parks, the palace moat and rivers. They allow views of a raw Tokyo like a roller coaster plugged into the city. It is possible to glimpse parts of the mechanisms which support the huge logistics of transporting people and large numbers of physical objects. What Le Corbusier looked at with the planning for Algiers and Marinetti proposed in his Futurist work, becomes actually constructed here, and linked to make an extensive network.
In Japan, the 20 or so years between 1966 and 1988 saw the number of cars increase by 6.25

川等地点为中心，如同在城市里建起一条过山车轨道。在这条轨道上不仅能看清东京的地形和建筑群之间的关系，还能一窥这部每天运输大量人员和物资的机械的真容。勒·柯布西耶为阿尔及尔做的城市规划，以及马里内蒂的《未来主义宣言》都预见了以高速交通工具为基础的动态城市空间，而这种设想现在以高速交通网络的形式被完全实现了。与之相应，日本每年在物流上的花费要占国民生产总值（GNP）的 10%。特别是最近的 20 年，快递系统急剧扩张，目前处理的包裹数量已经超过 16 亿个。这个系统由物流中心、配送站点、卡车、货车、高速公路、无线通信和互联网等多种工具组成快递网络，取代了街角的便利店、蔬菜摊和洗衣店。从 1966 年至 1988 年大约 20 年间，人均汽车拥有量增长到了原来的 6.25 倍，而公路的长度却只延长到了原来的 1.12 倍。如果把城市比作生物，交通网就如同把各器官联系在一起的血管，其中流动的血液量（汽车）急剧增加，而血管（公路）的数量却几乎没有变化。血管即将破裂，用雷德朋原则（步道、车道分离）这样的规划性手段动手术也无济于事。血管必然开始入侵各个器官。一下子被缝合入各个内脏的血管和为其输送氧气的增压器完全暴露在内脏中工作。也就是说，道路被修到了建筑物里，跟着建筑物的结构绕来绕去。与车分隔的舒适步行空间已成过去，取而代之的是一种行人、电车、汽车混杂的紧张刺激的交通空间。这个新原则让两地之间的移动更加便捷，这可能是人们第一次接受将交通空间和人居空间融合这一“物流城市”的规则（？）。

物流网络的发展为城市带来了重大的革新，成为决定真实城市形态的重要因素。“东京制造”中就有几座反映了这种革新的建筑物。

“16. 汽车塔”“18. 物流综合体”“23. 配送螺旋”“25. 出租车楼”“26. 货车塔”“59. 果蔬小镇”“61.TRC（东京物流中心）”都是和道路交通有着紧密联系的建筑物。它们是物流公司、汽车制造商、出租车公司或蔬果市场的公司房屋兼停车场。车就像人一样进入建筑，通过斜坡到达不同的楼层。通过这

times, but the length of road only increased by 1.12 times. Transport infrastructure which allows physical distributions is like the blood vessel system connecting organs for the city. For Tokyo, which is on the brink of sclerosis, it is apparent that Radburn's planning theory of separating cars and pedestrians is completely impractical. By necessity, cars have entered into the realms of people. Previously, the criteria for comfort was seen as separation of the traffic space of cars and trains, and human pace. But now a mixture between them occurs, which has become accepted as a new norm. The contemporary principle is about being able to clearly imagine how to connect to any particular location, and have easy access to the goal. Probably, the people of Tokyo have accepted the mixture of traffic space and human space, almost as a version of an 'urban regulation.'

In this way, traffic space has intruded into architecture in order to allow the execution of the highly developed goods transportation systems. The organ which is abruptly connected to the artery and the pump which pushes oxygen to that artery is exposed and pulsating in the urban environment as a single entity. The architectural design of this revealed figure is also exciting. The development of the logistics network has affected a huge change, and is one of the defining elements which has shaped the form of the city. Several of the buildings of Made in Tokyo reflect this kind of change and beautifully condense the dynamism of this city.
Examples of buildings united with the expressway:
13. highway department store
Examples of the integration of architectural floor plate and roadway:
16. car tower

样的做法，也许不下车就能够满足人的所有需求。在这里，“先下车再进门”的既定秩序被打破了。建筑物内部的坡道和连接平台都被设计成螺旋状或栅状等独特的形状，车道被立体地折叠到建筑物中，为城市弥补了道路空间上的不足。建筑物内部被分隔为停车、维修、包裹分流等区域，好像是城市整体的物流循环中的一个冷凝器。

与此相对，在城市整体的物流循环中也有一些在汽车通过时进行维修、补给的设施。“10. 首都高速公路巡逻站”通过坡道把巡逻车停车场及其管理设施和首都高速公路连接起来，还在停车场上修建了员工宿舍，是一座拥有直接联结首都高速公路特权的建筑物。“28. 跃层式加油站”是立交桥上下两段分别连接的两个加油站，碰巧堆叠在立交桥旁的同一座建筑物里，上方还建了石油公司的办公室。“65. 立交枢纽航站楼”是一座被通往机场的几条高速公路包围的航站楼，混在普通公路和高速公路形成的多层结构里，走在街上几乎辨认不出建筑的外围轮廓。

汽车数量的激增意味着停车场数量的激增。停车场其实就是汽车的仓库，人口密度大的繁华地段尤其需要停车场。地价高昂，汽车却不能折叠，十分占地方。为了解决这个矛盾，就需要积极地兴建多层复合式建筑物。“55. 汽车办公楼”的地下一层、一层和二层都是餐厅，三层及以上是停车场。但建筑物本身并没有斜坡等特别的要素，而是采用幕墙式的构架（但没有使用玻璃），完美地融入周围的建筑群之中。圆筒状的电梯垂直贯通七层地板的中心。因为面积很小，无法修建长的通道，所以电梯一边上升一边旋转，呈放射状把车运往各个楼层，像一个汽车轮盘。车只能直进直出电梯，以电梯为中心呈放射状分布于各个楼层。“12. 公园停车场”为了不让停车场这个“城市的内脏”外露，“公园停车场”躲进了建在人工地基上的公园底下。上层公园里的树木以榉树为主，粗壮茂盛，人们很难想到下层还藏着一个停车场，这些榉树甚至成了一些流浪者临时搭建的小屋的屋顶。这是车的 park（ing）（停车场）和人的 park（公园）的层叠。

这些深入到交通系统里的建筑物，即使从设计的角度看已是完成品，也绝不会变成一些孤立的纪念碑。因为正是这些分散在东京各处的建筑物互相联结，一同构成了这座庞大的物流城市。“条条大路通物流城市”这种和不在眼前的东西发生联系的状态，也可以说是与邻接性相反的状态，我们暂时把它命名为“远隔性”。

18. distribution complex
23. delivery spiral
25. taxi building
26. truck tower
59. vegetable town
61. tokyo dispersal centre
Examples of transit buildings:
10. expressway patrol building
28. double layer petrol station
65. airport junction
Examples of special car parking:
12. park on park
55. car parking office
These individual buildings can never become monuments. The reason is that they are all dispersed throughout Tokyo, each forming just one part of the circulation system and only all together making up what we can call a logistical urbanity. 'All roads lead to logistical urbanity.' So, now please delve into the logistical urbanity and experience it for yourself!

SPORTIVE

Let's imagine a situation of being thrown into various environments with a single board under our feet. If that happens to be the big wave of the sea, then we are surfing. If that happens to be the slope of a snowy mountain, then we are snowboarding. And if that happens to be on the asphalt of the street,

运动场

想象一下把一块板放在两脚之下，进入各种不同的环境的状态。如果身处波涛汹涌的大海，那就成了冲浪；如果身处有坡度的雪原，那就成了滑雪；如果身处铺了柏油的城市街道，那就成了在玩装着滑轮的滑板。通过同样的一块板，我们可以观察不同的环境，并且察觉到行为实际上就是“与环境信息相互作用”。詹姆斯·吉布森认为可供性是“一种环境性质，被界定为环境与动物（人）之间的关系”，而现在的城市运动也以通过身体运动发现环境所具有的“价值”与“意义”为基础。发现城市环境中的空白剩余，通过身体行为将其表面变为运动场所，也就揭示了城市的环境要素之前从未有过的一面。身体与环境的联结越是紧密，纳入的环境要素越多，运动就越具有野生的性质。

“东京制造”的建筑物通过堪称典范的精巧构思将宽阔的屋顶、用笼子隔出的空地、盆景式的地形转化成各式各样的运动场所。

“1. 仓库球场”是在 JR 公司货仓的屋顶搭建的网球场。在耐克的广告中，西武狮棒球队的松井选手和其他运动项目的著名选手一起乘坐某体育百货大楼的电梯，来到了一片足球场（另一个世界）。而这座建筑物也一样：搭上和仓库完全无关的电梯，就能到达一片绿色的空中球场。从近旁的首都高速公路眺望，被夜间照明设施照亮的屋顶就像是另一个世界。在这个世界里也要“Just Do It”！

“27. 立交球场”和“66. 运动桥”都在道路这一土木构筑物的空隙里建了网球场。温布尔登网球赛总是请观众保持安静，在一片肃静中竞技；而来这里打球的人，周围与下方是往来不绝的车辆。

完美地融入居住区的“67. 运动者动物园”利用深绿色的斜坡打造了集高尔夫球练习场、室内外网球场、带有室内高尔夫球场的俱乐部会所为一体的复合建筑群。作为人科动物的运动者们在各种挡网和围栏里打球、跑跳，看起来很好笑。目黑川沿岸的“14. 高尔夫出租车站”则截取了出租车公司办公室和停车场上方的空间，用网围出一条盆景式的高尔夫球道。用以阻挡来球的网笼形态多变，可以被设计成各种不同的形状和大小，而在这座建筑物中，网笼下方的空间被有效地用作了出租车公司的停车场。更巧妙利用网笼的例子是“54. 皇家高尔夫公寓”。它把高尔夫球练习场的网笼放置在了被公寓楼、家庭餐厅、停车场、俱乐部围住的剩余空间里。公寓的走廊紧挨网笼的边缘，站在走廊上就可以呼唤挥汗击球的爸爸。这座集运动、住、食为一体的建筑物让爸爸们足不出户就能享受周末。

then we have added wheels and are skateboarding. The common vehicle of a single board helps us to observe differing environments and lets us perceive the reality of action as 'interaction with environmental information'. According to James Gibson, 'affordance' is the nature of the environment as defined by its relation to living beings. Contemporary urban sports are also based on the discovery of various values and meanings furnished in the environment through the movement of the body. By finding residual spaces inside the closely packed urban field, and using human action to turn those surfaces into sports fields, the elements of the city gain a whole new appearance. The body and the environment become inextricable, and the more urban elements are dragged into the sports scene, the more that particular sport goes wild. In their capacity to turn roof tops, caged-in skies and boxed in geography into broad sports fields, the buildings of Made in Tokyo exhibit the kind of fine play which should be enshrined.

Examples of boxed in geography:
27. interchange court
66. sports bridge
Examples of caged-in skies and seas:
14. golf taxi building
42. diving tower
54. royal golf apartments

“42. 潜水塔”是一座微型的蓝色潜水练习设施，像是用水罐截取了电影《碧海蓝天》的布景，改造后的工厂水罐内部犹如大海。从水下的窗户看到的东京，大概比海底的蓝色还要美吧?

“30. 污水球场”和“36. 双子污水花园”为了让扰民的污水处理厂与城市共存，把顶部的剩余空间改造成了运动场。虽然上下两层在功能和管理方面都毫无关系，但对于周边居民来说把如同“皮鞭”的污水处理厂和如同“饴糖”的运动场与公园如此组合起来（形成“皮鞭加饴糖”的复合体），颇能显示出行政人员的老奸巨猾。尤其是位于落合，能眺望新宿高层建筑群的“30. 污水球场”，于昭和三十九年（1964年）东京奥运会期间建成。当时它是世界上第一座把顶部改造为公园的污水处理厂，堪称“皮鞭加饴糖”复合体的先驱。加上夜间照明设施，污水处理厂变身为“梦幻之地”，让人几乎忘记它是城市的一个消化器官。

位于葛饰区龟户的“31. 净水球场”是层叠在净水厂上方的棒球场，四周被邻接的集合住宅环绕，所以也可以被看作公团住宅的中庭。各家各户就像体育场里单独的座位。这个例子不仅是建筑物的层叠，而且还有周围环境的参与，形成了有意义的统一体。

今后，东京的城市环境一定会催生出新的体育运动吧。“读卖巨人队签下排在选秀第一顺位的击球练习场球员”“网球对墙击球世界锦标赛决赛”“BMX 小轮车与滑板跨界对决”……大概不久以后，报纸的体育版上就会出现这些标题了。

副产品

今天，东京被建筑物、高速公路、铁路等人造物塞满了。虽曾两度遭遇大火，但在现代化的进程中，许多物理结构物因各种各样的目的而兴建了起来。不过，城市里的物理结构物不总是完全服务于它们的目的，即人类活动，通常会有一些显得多余的部分。例如，高速公路的目的是让车更快地通行，但在用地紧张的东京，它们却大多被设计成高架桥，于是就形成了高架桥下多余的长条形空间。仓库是用于保存大量物品的空间，由此产生了墙壁、屋顶这些巨大的外表面。独栋住宅的样式决定了建筑物之间的空隙会成为不可避免的死角，这样的死角无处不在。这些空间是在建筑物达成自身目的的过程中产生的副产品。如果数量不多，倒也微不足道；然而，一旦数量和密度超过某个界限，它们便开始参与塑造城市空间。城市的副产品不只是垃圾和犯罪。在今日的东京，

67. sportsman zoo
Examples of 'candy-whip' rooftops:
30. sewage courts
36. twin deluxe sewerage gardens
Examples of unexpected rooftops:
1. warehouse court
31. supply water courts
Probably, many more different types of sports inspired by the urban environment will continue to emerge in this city. It may not be too far away that we see headlines such as 'Batting Centre Player Hits The Big Legue!', 'Wall-Tennis World Cup Finals', 'BMX vs Skateboard Real Fight!'

BY-PRODUCT

Today, Tokyo is completely covered by constructions: buildings, expressways, railways; even though the city has twice been burnt to the ground. During the process of modernisation, many physical constructions were built for various reasons. But the city's physical constructions do not entirely serve the equation of aim = people's lives. Mostly, there has been produced a surplus, over and above the raw requirements of the aim. For example, the aim of the expressways is to let cars run as fast as possible, but because the acquisition of land is very difficult in Tokyo, most of the expressways have been raised off the ground, which produces enormous amounts of under- infrastructure space. The aim of warehouses is to store as much as possible, but they also result in huge surfaces of wall and roof planes. The form and layout of the detached house type

这样的现代化的副产品和主产品杂乱无章地混杂在一起。如何处理这些计划之外的副产品与原初的建筑规划之间的关系呢?

思考这个问题大致有两个方向。第一个方向是思考不产生副产品的方法，另一个方向是思考积极“使用”副产品的方法。尤其是在如何把已经存在的大量副产品转变为建设城市的资源这一点上，后者意义非凡。“东京制造”的建筑物中也有若干活用副产品的范例，我们从中可以学到许多。

“运动场”一节提到过的“1. 仓库球场”“30. 污水球场”“31. 净水球场”都将不得已产生的楼顶空间改造成了运动场，同样利用楼顶空间的例子还有“68. 直升机仓库”“29. 超级驾校”“33. 阿美横町空中寺院”。它们把楼顶看作另外一片地面，如果这些地面能被连成网络，那么城市空间将变得有趣得多。

相机店可以说是新宿的一大标志，店外的墙面被无数霓虹灯和广告图片覆盖。位于新宿东口站前的“6. 霓虹灯楼”的墙面一层写着商品名称，二层以上则写满了商品种类的名称，屋顶部分则安装了展示店名的霓虹灯，只有小窗户空着。不同的店铺重复使用着相同的形式，如果不注意，根本不会留意到隔壁大楼里的是另一家店。这种广告形式把“绝不浪费任何机会”的商业思维完美地展现在了城市空间里，同时这也说明市民们乐于接受这种形式的广告。将文字之类的符号放大到与建筑物差不多大的尺寸，并且使用电力装置使其闪烁跳动起来，城市空间的样子会因此改变。虽然已有几位建筑师讨论过这种做法，把它看作一种现代建筑的表现形式，不过文本与图像的融合今后势必成为更多人研究的课题。“35. 铁道博物馆”，即东武线车站下方的东部铁道交通博物馆，把高架铁路下的细长空间用于展示铁道车辆，正在使用中的列车就在已成展品的铁道上方行驶，这一点既奇妙又合理。虽然万世桥的国立交通博物馆才是交通博物馆的正统，但它更改了砖造的旧万世桥站的用途，所以万世桥的轨道现在只用于调度待检列车。位于京王线高尾站旁高架桥下的“40. 蜈蚣住宅”在约 300 米长的区域内容纳了 43 栋独立住宅，用作铁路公司的员工宿舍。住宅被高架桥的桥墩均匀切分成同样的规模，样式也完全一致。为避免传导振动，每一户在结构上完全独立，所以这里不像是被强占的民宅那样充满生活气息。虽然各户分居，但都处在同一屋檐，即公司建造的高架桥下。

除此之外，“2. 电器走廊”“3. 高速百货

makes gaps between house and house become a mass production of dead space. This kind of space is like a byproduct. Whilst its density stays minimal, its meaning is insubstantial, but once it reaches a certain quantity and density, it starts to make its own impression on the urban space. Urban by-products are not only criminal activity and rubbish heaps. Contemporary Tokyo's situation is a crazy mixture of the main-products and by-products of modernisation. How can we feed back the birth of such unexpected products into the urban planning system?
There are two main threads to thinking about this issue. One is to try to think of a method where by -products do not ensue. Another is to try to think of a method where by-products can be actively utilised. In particular, when we consider that the city already abounds with ignored and undefined surplus spaces, it seems appropriate to try to rediscover these as a positive resource. It becomes important to think of a way to actually use them. The buildings of Made in Tokyo have many good examples of by-product utilisation, and there is much we can learn from them.
Examples of rooftop utilisation:
15. warehouse court
29. super car school
30. sewage courts
33. ameyoko flying temple
Examples of wall plane utilisation:
6. neon building
7. pachinko cathedral

商店”这两个在“跨界格斗”一节中提到过的例子也都利用了交通结构物的副产品。

基本上可以断定，这些随处可见的副产品有成为城市建设资源的潜力。最后举一个“东京制造”里副产品成为独特城市景观的例子，不过这次不是在高架桥下。首都高速公路在立交枢纽处一定会形成螺旋状的匝道，“27. 立交球场”把这些匝道中央的空地改造成了网球场。在被螺旋状匝道和白噪音包围的球场打球，仿佛置身于美国网球公开赛观众大声欢呼的现场。

此外，还有通过填埋，在跨越运河的桥下方造出地下空间，并将其改造成电影院的“4. 电影桥”；建在私营铁路地下化改造之后形成的地上长条形空地上的“60. 松茸形车站楼”。它们都利用了独特的副产品，这些副产品是城市改造的活生生的见证。

城市规划的副产品，比如楼顶、墙面、高架桥下方的空间，以及我们没有举例的、大量出现于居住区的住宅与住宅之间的空间，它们都没有预设的用途，只是随处可见的空白。如今，人们由于空白恐惧症而一心只想着填满这些空隙，孤立地思考它们的意义，但这样也就忽视了它们的多样性。这些空白是过于密集的城市空间的喘息之所（OFF），可以不断变换用途。这些副产品，即空白及其集合，将成为今后城市建设的重要工具。

城市居住

空气清新、绿意盎然、环境安静、阳光直接透过朝南的窗户洒满整个宽敞的房间，如果想在东京这样的巨大城市带找到这么一个理想的舒适居所，要么逃往郊外，要么就得付出高昂的成本。为了让如此高密度的居住环境变得更加舒适，建筑师和城市规划师们设想了“塔状城市”“东京湾城市”等让田园和城市共存的乌托邦，但这些乌托邦未必能成为现实。不过，在现实的城市里，努力适应环境以克服城市病，在此过程中获得新型免疫的建筑实践却比比皆是。

目前，东京 23 区中有多达 68% 的人从事第三产业。从明治时期以来，对于能在如此短的时间里实现高密度城市化的劳动者和经营者来说，生产、流通的高效率就是最大的善。如果以这个标准来评价居住这件事，就会得出“工作生活无限接近”这样简单直接但说出口

8. sex building
Examples of under-infrastructure space utilisation:
2. electric passage
3. high way department store
35. rail museum
40. centipede housing
Other examples:
4. cine-bridge
27. interchange court
60. sprouting building

These urban by-products; rooftops, wall planes, under-infrastructure, and the abundant gaps between houses; are void spaces which generally avoid a pre-defined use being allocated to them. At the moment, the usual attitude is a kind of void-phobia which tries to paint over all trace of such spaces and they are considered only individually so that the meaning of their multiplicity is lost. However, these voids can become a breathing space within the over-dense urban environment of Tokyo, and can be recycled into a completely new use – the equation of by-product = void and its assemblies can become a tool for a future worm's eye view urban planning.

URBAN DWELLING

Within the extreme density of Tokyo, real estate ideals – clean air, lots of greenery, no noise pollution, big rooms, plentiful sunlight from large south facing windows – are like a far away dream. Trying to approach such ideals entails huge cost; enormous monetary expense, or being pushed out to extremely distant suburbs. The uncontrollable spread of this situation is like a virus. Modern architects and city planners tried to counterattack the effects of this highly dense living environment by taking utopian ideas combining the rural and urban, and trying

需要勇气的结论。但正如“日本制造”（比如随身听和方便面）极致的功能主义里有极致的美感，“东京制造”观察到的城市居住的形式也为我们如何在未来的城市环境里生活提示了新的方向。

“15. 混凝土公寓”是混凝土工厂和司机宿舍的复合体。为了按时把混凝土送到建筑工地，司机干脆和生产设备“同吃同住”。人的血肉之躯融入了城市巨大的建设系统里。“22. 渣土公寓”和“18. 物流综合体”也是如此，前者是土木工程公司在堆放建材的架空层上建造的办公室和员工宿舍，后者集物流公司的车库、配送站、办公室和员工宿舍于一体。邻接大井赛马场的“17. 马公寓”，一层是马厩，二、三层是驯马师宿舍，就像有谁颁布了要善待马的法令一样，人和马平等地生活在一起，33座相同样式的建筑物齐整地排成一列。从经济的角度看，赚得更多的显然是马，马说不定才是主人。以上都是“工作生活无限接近”的例子。

在“50. 挡土墙公寓”中，从建筑物延伸出的井字梁直接成为挡土墙的一部分，暴露在走廊的空间中。井字梁从中段开始被涂上白漆，虽然公寓的界限到此为止，但从力学的角度看，这是整体上不可分割的结构物。净妙寺川沿岸的“64. 水坝公寓”外的地面平时是广场，下大雨河川涨水时则会变成池塘。这是为防止河川泛滥而修筑的蓄水池，在雨势变小时水池里的水就会被放出。正常状态和紧急状态时的模样十分不同。像这样同时具备日常建筑和治水结构物功能的复合体，基本都是以应急功能为前提，但平时却能呈现出完全不同的功能，十分有趣。“51. 桥洞里的家”是一家儿童收容所，位于通往青山墓地的高架桥下，高架桥的配色是勒·柯布西耶风格的。虽然地处青山这样一个时尚繁华的街区，但他们悄无声息地居住在土木构筑物之间。以上是紧邻土木构筑物而居的例子。

“自动尺度”一节提到的“13. 公交住宅区”和“副产品”一节提到的“40. 蜈蚣住宅”都是由和居住本身毫无关系的原因决定了居住单元的尺寸。所谓设计，更多时候是以人为本，让物适应人；但在这两个例子里，反而是让人适应了物。如此倒错，如此幽默。

动画《机动战士高达》里有一群出场角色被称为“New Type”，他们是在宇宙时代的新环境中进化的人类。虽然现在还不是动画里

to project them into towers or onto Tokyo Bay. However, amongst contemporary urban dwellings, there exists an immunised type , which has adapted to the actuality of the current condition in its own attempt to overcome the urban disease.

In Tokyo's 23 wards, 68% of employees are working in the service sector. The workers and managers propelled the speed and intensity of urbanisation, and regard the efficiency of production and logistics as a resulting virtue. If dwelling is also thrown into the same framework of this criteria, then a very simple yet courageous thesis of 'hyper-closeness of home and x' takes effect. Like the absolute beauty seen in overwhelming functionalism of products made in Japan, such as the walkman and cup noodles, the format of urban dwellings observed in the buildings of Made in Tokyo can guide us towards new ways of living in the future urban environment.

Examples of living with work:
15. nama-con apartment house
17. horse apartment house
18. distribution complex
22. sand apartment house

Examples of living with civil engineering:
50. retaining wall apartments
51. bridge home
64. dam housing

Examples of living with given measurements:
13. bus housing
40. centipede housing

Twenty years ago, there was a TV animation series which attracted popularity with the concept 'new type'. The 'new type' meant an evolved human who had adapted to the new environment of the space age. Tokyo is not in the space age at the moment, but through looking at these contemporary dwellers,

描述的宇宙时代，但从居住在以上建筑物里的现代人身上，我们不是已经看到适应现实城市的各种各样的环境，并努力生活在其中的“New Type”的身影了吗?

作为建筑物的机器

城市是我们生活的地方。但如果把视角拉远，观察城市的整体，我们会发现城市是一个不断重复地进行生产和消费的类似机器或生物的有机组织。还是沿用这个比喻，城市也需要动力器官、传输器官、排泄器官和储存器官等。在东京这样的巨大城市带，各种器官负荷巨大，为了应对这些负荷，器官本身也变得很大。这里说的器官可以理解为各种各样的城市设施，特别是在东京湾一带，集中分布着符合上述比喻的设施。发电站、垃圾焚烧厂、污水处理厂及其他类似基础设施的建筑物就是典型代表。在现代主义建筑的历史中，它们被认为是工厂建筑，所以外观通常没什么不同寻常之处（特别是近来，垃圾焚烧厂还被规整地集中到了一起）。不过，在新出现的与基础设施相关的建筑物里，我们却见到了奇怪的形态——与建筑一样大的机器或工具。勒·柯布西耶曾提出“作为机器的住宅”，在“东京制造”中，我们见到了它的反面——“作为建筑物的机器”，即把建筑比喻为机器。

“19. 空调楼”是为羽田机场周边区域提供冷暖气的设施，由热交换器、冷却塔、储水罐和众多管道共同构成一个地上的建筑整体。它就像是机场所有相关设施的大空调，因为要负责附近整片区域的冷暖气供给，所以足有七层楼高。对比同一路段的其他建筑物，其巨大的楼体和高科技的外形十分显眼。要是没有横铺在建筑物正面的外露管道，它就能毫不违和地融入城市景观。“62. 冷冻园区建筑群”这台东京的大冰箱不是在工厂而是在现场制造的。如果你家冰箱里的东西出自超市里的冰箱，那么超市冰箱里的东西一定都属于一个更大的冰箱。这些相当于你家冰箱祖辈的大冰箱们平铺开来，占据外部空间，再配上卸货平台和办公室，加上在其间活动的货车和人，就呈现出城市的或建筑的面貌。原则上，把这些冰箱倾斜过来，就成了室内滑雪场。像这样，在整座城市的范围中搜集这些日常生活里琐碎的细节，细节的规模就会扩大，逐渐超过人的规模。这些“作为建筑物的机器”也就可以被看作城市的各种家用电器。

“48. 起重机架”是塔式起重机的床铺。在泡沫经济时期，这些塔式起重机丰富了东京的天际线。泡沫经济时期后，塔式起重机都去了哪里呢？没想到是以这种方式堆叠起来，呈现出临时建筑物的面貌。构成起重机的桁架在

living in these buildings and overcoming various contemporary urban difficulties, I feel I can see the figure of the 'new type.'

MACHINE AS BUILDING

The city is our place for living. But if we zoom back, it can be seen as an organic structure like a machine or a creature breathing production and consumption. Although it is an analogy, if we think this way , then the city starts to need organs for power, transportation, storage, discharge. In the case of a megalopolis like Tokyo, the load on such organs is enormous, and the organs required to look after them must also become huge. In reading these organs as built facilities, they can be readily seen in the area around Tokyo bay. The most typical examples of these are almost-infrastructures such as power plants, garbage incinerators and sewerage plants. Generally , they have been regarded as types of industrial building; because of Modern architectural history's interest in them , they do not seem particularly strange to us now. However, in the latest infrastructure related facilities of Tokyo, bizarre buildings which are really like large machines or instruments have appeared. These are what we have focused on for Made in Tokyo. According to Le Corbusier, 'the house is a machine for living in,' but according to Made in Tokyo, the reverse 'machine as building' is also true. In other words, architecture as an analogy for machinery is possible.

结构上完全独立，看起来就像层层重叠的建筑。希望这座临时建筑（?）被拆掉的时候，经济状况会有好转吧。“43. 底盘公寓”则是拖车底盘的床铺。用于海运的集装箱被设计成 6 米和 12 米两种规格，再考虑到对车体的有效利用，底盘的数量和种类是拖车本身的两三倍。集装箱数量的增加会带来相应规格底盘数量的增加，底盘公寓的构想是一个可以收纳新增底盘的合理系统。“作为建筑物的机器”就像是分散在城市各处的工具们的收纳库。

“37. 增殖滑道楼”的一层是游戏厅、停车场和餐厅，楼顶开设了泳池。螺旋滑道是泳池的一大亮点。为了追求更加刺激的滑行体验，滑道每年都在扩建，足有八层楼之高。螺旋滑道以奇异的姿态出现在居住区上空，被孩子们的欢呼声和年轻女性的嬉闹声环绕。它应用了常用于船舶内部的高级配管技术，与物理或化学设备无异。产业技术被用于为人创造娱乐体验。“69. 洗车露台”的一层是自动洗车区，二层是手动洗车区，两种模式（自动、手动）的洗车区域形成层叠结构。为了达到洗车的目的，人的身体也变成了一个零件。“作为建筑物的机器”直接将人的身体与净水和污水管道等基础设施相连。

为缓解西武新宿线道口引发的环状 8 号线拥堵而修建的井荻隧道（全长 1253 米，4 车道），在其上建了两座通风塔（直径 7 米，高 35 米），用于排除汽车尾气。这两座通风塔和隧道监控室、消防室、仓库等组成的复合体就是“38. 通风方尖碑”。它们对称分布于隧道的两侧，因为装有用于排气的烟囱，所以比周围的其他建筑都要高。在道路中央竖起无窗的黑色圆筒状结构物，对于远处的司机来说也是极易辨认的视线落点，让人以为是巴黎广场上的方尖碑。方尖碑似乎代表着对这个区域的祝福，但杉并区并没有赢得什么战争。难道是为了纪念交通拥堵、汽车尾气这些城市的必要之恶吗？东京都内散见的其他隧道的通风塔都明显比周围的建筑物大，其巨石般的形态没有什么生活气息，人们也不知道它们的真正功能。充满神秘色彩，和周围的环境格格不入，十分符合纪念性建筑的特征。“作为建筑物的机器”成了偶然的纪念碑。

不止东京，其他大城市里与基础设施相关的建筑物，虽然为人们的生活提供服务，但是经常被当成麻烦的东西。这些设施往往和机器一样有非常明确的用途，随着城市的扩大，它们的体积也不断膨胀。与此相对，普通建筑却转向另一种截然不同的价值观，追求多功能性。然而，普通建筑构成的平凡的城市织体和反映了日常生活刚需的奇形怪状的大型基础设施其实是互相依存的关系。“作为建筑物的机器”

Examples of electric appliances for the city:
19. air-con building
62. cool room estate
Examples of urban tool garages:
43. chassis apartment
44. TTT (lego office)
48. crane shelves
Examples of direct human plug into waterworks:
37. proliferating water slides
69. carwash terrace
Examples of infrastructural monuments:
38. ventilator obelisk

In a metropolis such as Tokyo, infrastructure related facilities are often seen as troublesome, despite the fact that they serve the populace. They have clear aims, as machines do, and gain obesity in direct relation to the extent of urban expansion. In contrast, normal architecture must adjust itself to multiple influencing forces, and cannot afford to stand with a single purpose like these facilities. But the banal fabric of normal urbanity is actually co-dependent with the figures of the huge and bizarre infrastructural facilities, which are merely a reflection of inevitable daily needs. The structure of the machine-as-building has had a scapegoat symbolic value attached to it, which can be replaced with more positive values, providing a unique opportunity for urban proposals.

成了具有象征意义的替罪羊。如果整个社会能以更积极的态度观察它们，“作为建筑物的机器”也许将成为改造城市空间的独特契机。

城市生态系统

东京是由住宅、办公楼、工厂等建筑物，公路、铁路等交通设施和海港、公园等土木构筑物凝缩而成的集合体。这样的景观没有视觉上的统一，通常被认为处于“混沌”或者接近“白噪音”的状态。但这样的解释以构造性的与符号性的视觉秩序为前提，如果改变这个前提，就会产生完全不同的解释。虽然现在的东京被认为处于混沌状态，但它以自身特有的有趣方式运转着。就像热带雨林一样，既缺乏可见性，也没有统一的形态，但不同种类的生物一边构筑适合自己的环境世界，一边实现共存。因此，如果暂时放弃机械论与符号论的解释，采用在完整的生长环境中观察生物个体的环境生态论视角，就能从东京的风景中发现若干有意义的环境统一体吧。通过日常的行走，我们能发现从城市上空俯瞰时无法看到的景象：一个由人、物体的流动、环境和时间等复杂要素交织组成的城市居民的小剧场，也可以说是一个微型的城市生态系统。据此便可以想象由这些东西聚集起来形成的、作为一个巨大生态系统的都市整体。这些连续进行着各式活动的舞台，极少只由单一的建筑构成，通常都纳入了建筑周围的环境要素。在这样的情况下，建筑物的完整性、建筑 / 土木工程之间的区分都显得不那么重要了。

“东京制造”收录的建筑中不乏精彩的建筑案例。它们有如微型的生态系统，既直接反映了城市居民们各式各样的立场、趣味和城市的系统构造，又展示了建筑周围的环境统一体。这些建筑就像是一幕幕包含了戏谑、幽默和悲悯的都市短剧。

“7. 柏青哥教堂”乍一看是一座独立的建筑物，实际上是由三座邻接的建筑物组成的。居中的尖塔型柏青哥店被两座主要由高利贷公司占据的铅笔楼包夹，外形神似巴黎圣母院。亚克力板和霓虹灯组成的招牌代替《圣经》的故事占据了大楼的外立面，一入夜就熠熠生辉。金钱被奉为上帝，在柏青哥机、老虎机上赢钱就得到了救赎，而输钱就要到两侧的高利贷公司借钱忏悔，再回到机器前继续赌。城市生态系统中可怖的一面就这样被对称地打包到一起。这个微型生态系统由几座独立的建筑物组成，它们在形态上偶然地构成了一个整体。

“46. 公寓山寺”位于一块搭建在斜坡上的人工地基上，要前往寺院参拜，必须经过公

URBAN ECOLOGY

Tokyo is an agglomeration of buildings, traffic infrastructure, civil engineering. Its landscape is said to lack visual control and is popularly thought of as chaotic or as ‘white noise.’ However, this kind of interpretation is based on mechanistic theory and semiotic systems. So, if we change this premise, a totally different interpretation of the city should be possible. Actually, despite these claims of chaos, Tokyo is interesting in its own way of functioning. It resembles the unstructured forms of the rainforest, within which there is in fact many types of creatures co-existing, whilst each constructing their own world. This is ecology, which understands the creature itself in relation to its living environment. If we stop using the metaphors of mechanistics and semiology and start using the metaphor of ecology, then it should be possible to discover layer upon layer of meaningful environmental unities, even within the landscape of Tokyo. This is a complex intertwining of people, the flow of things, elements of the environment and time; something which can never be obtained by the bird’s eye view. Through walking around the reality of everyday life, we can start to see an urban micro-ecosystem, or theatre of urban dwellers. Then, we can also start to form an image of a city accumulating from these variable

寓内部的楼梯和公寓楼顶。走过连接楼顶的桥，会来到用防水布代替粗砾石铺设地面的寺院前庭，寺院两侧的灯笼被鸽子小屋（通风设备露出屋顶部分的别称，并不是饲养鸽子的小屋）替代。“41. 汽车村”囊括了汽车销售、保险、涂装和维修等业务，位于中央的转车台被停车塔环绕。远远看去，这里就像一座超微型底特律汽车城。附近的练马车检场与之配套，可以满足附近车主的所有需求。在这个例子里，几座不同的建筑物形成整体的原因不是形态一致，而是功能互补。

位于高圆寺站附近，正对着环状 7 号线的“24. 澡堂旅游楼”让单身的人们在同一座建筑物里就能体验“自助洗衣店→澡堂→桑拿房→便利店→啤酒→打嗝”这样解放身心的夜生活。“33. 阿美横町空中寺院”位于阿美横町商店街上方的人工地基上，所以沿街的商店也兼作参拜道边的货摊。在商店店员中气十足的叫卖声中，寺院在商业区的上空忽然出现，让人顿生一种神灵必然会保佑自己的预感。这是互有关联的文化娱乐设施被打包进同一结构体的例子。

在“25. 出租车楼”里，连接外部公路的车道直接穿过建筑物，办公室、休息室等属于司机的空间和修理厂、车库等属于车的空间被打包放置在同一座建筑物里，真可谓“同床异梦”。小憩、出车、再小憩，一天中每时每刻都有车从这艘母舰离开，朝城市进发。在“57. 汽车百货商店”里，车主将车直接开进二层的停车场。他们可以一边在楼上的汽车用品商店买东西，一边让爱车在一层保养，还可以到最顶层打上一局保龄球或电子游戏。汽车百货商店就是一处立体的中途停车服务站点。以上是以汽车相关服务为主，把其他派生服务一并打包进同一座建筑物的例子。

“47. 吸血公园”的原址是秋叶原电器街和 JR 高架桥之间的果蔬市场。果蔬市场被拆除后，电器街大楼的背面就暴露了出来，设计者在柏油路上画出三对三篮球的场地和滑板赛道，吸引了大批年轻人。不过，公园的角落悄悄设置了一个献血站，换个角度看，整个公园就像是吸收青春血液的装置。一间小屋的存在，改变了人们看待整个公园的方式。同样的例子还有在“宠物尺寸”一节里提及的“52. 宅地农场”。“70. 射击墓地”由朝霞自卫队的射击练习场和邻接的隔土堤而建的墓地组成。从墓地只能看到隆起的土堤（装有雷达），却能听到对面不时传来“砰！砰！”的枪声。射击练习场利用了古墓的地形，原本用作东京奥运会的比赛场地，但现在通常不向普通民众开放。这当然有安全方面的考虑，但是否也是为了防止民众对“射击练习场 + 墓地”这样不吉利的

happenings. This stage of connected action is brought into being by utilising every possible element from the surrounding environment. The completeness of any building and the categorical division between architecture and civil engineering becomes meaningless.
Within the collection of Made in Tokyo, we can see positive examples of micro ecosystems, directly reflecting the values, interests and social systems of various urban dwellers. They can be recognised as small urban episodes including jokes, humour, pathos.

Examples of physically separate buildings united by activity:
7. pachinko cathedral
41. vehicular village
46. apartment mountain temple
Examples of various amusements packaged into a single building:
24 . bath tour building
33. ameyoko flying temple
Examples of the agglomeration of cars and their derivatives:
25. taxi building
57. auto department store
Examples of adjacencies which inspire unexpected ecosystems:

组合浮想联翩呢？以上是城市的开放空间、空白空间与周边环境相互作用，形成意想不到的生态系统，激发想象力的例子。如果能多学习这些例子，现在大同小异的儿童公园也能变得多姿多彩吧。

像这样观察城市里的微型生态系统，就能感受到跨越建筑类别，无视建筑物的完整性，灵活多变地结合周围环境进行设计的乐趣。同时也能看到，受美学与建筑技术制约，束缚设计活动的做法有多么不自由。另一方面，这种不自由又反过来塑造了城市中滑稽的局部细节，让人感受到城市的“戏剧性”。总之，它们的共通之处是通过与邻接环境的互动，源源不断地生产出城市生活的乐趣。

虚拟用地

便利店和图书馆、美术馆、车站等其他建筑类型的决定性不同在于，它们是采用同样设计的小型建筑单元，以极高的密度分布在城市各处，形成网络。POS（销售时点情报系统）和基于这个系统的物流系统为便利店网络的运转提供支持。而且网络自身不断扩大，可以说形成了一个巨大的设施。这个软件 / 建筑的网络本身是肉眼不可见的。每家店铺只展现了整个网络的局部。在便利店的网络化经营战略里，虽然店铺的选址会是一个问题，但具体场所的个别特征不构成问题，可以通过货架的布局或者招牌的位置和设计来抹除场所的影响。室内设计的规划和布局这些几乎不受场所影响的部分才是便利店经营的重要因素。

便利店这种建筑类型的出现意味着“场所”这个概念不再是万能的。便利店为了融入各种环境，所以设计时会避开“场所”造成的问题。“场所”是现代主义成为主流以来，建筑师们解释建筑设计时理所当然的入手点，这么一想，便利店带来的变化就很惊人了。如果“场所”不再构成问题，那“用地”的概念也必须随之改变。各家便利店除了占据作为具体场所的用地（但在设计上不怎么重要），还占据着网络上的另一片“用地”。如果作为场所的用地是“现实”的，那么网络上的用地是“虚拟”的。虚拟用地和建筑设计之间存在怎样的关系，正是这一节要讨论的重要问题。“东京制造”包含以各种各样的网络为背景的建筑物，它们为讨

47. vampire park
52. residential farm
70. shooting graveyard

If we observe the micro eco-systems of the city in this kind of way, cross-categorical and innocent utilisation of the built environment can have plenty of fun, rather than being weighed down by the solemnity of a single building. We realise how unfree the limitations set on our activities by the control of aesthetics and construction technology are. On the other hand, it is these very same limitations which create the comical details of urban dramaturgy. In either case, the particular understanding of adjacency endlessly produces urban delight.

VIRTUAL SITE

The difference between convenience stores and other public building types such as libraries, museums and train stations is that there is an incredible number of them spread throughout the city, that they are made up of a small space of repeated design and that they have established a networking system. This network is supported by logistics systems which control informational and product exchange between merchants and their clients, which is known as the POS (Point of Sales) System. In one sense, we can say that it is the network itself that makes one huge public facility. But this kind of network as software-architecture is invisible. It is only ever the parts, each individual shop, which can actually be experienced. Therefore, in terms of the network's strategy, there may be an issue of exactly where to locate the outlets, but there is no problem in terms of site specificity in the design of each shop. For example, there is essentially no difference in the layout of merchandise or the arrangement of the signage within any network. In fact it is the sameness of the specifications of the

论的推进提供了素材。

集便利店、澡堂、桑拿房等功能于一体的"24.澡堂旅游楼"是一个极好的例子。正是由于便利店有占地面积小、不被场所束缚的特点，所以能融入各种各样的地方，创造各种各样的复合体。因为便利店的存在，这样的复合体，比起单独的澡堂和桑拿房提供了更多信息和物流支持。澡堂和桑拿房由此也被置于网络上的场所，即虚拟用地。正对着目白大道的"58.家庭餐厅三兄弟"是并排开在人工地基上的三家经营不同菜系的家庭餐厅，旁边还有一间高尔夫用品商店，而地基底下则是一片停车场。三家家庭餐厅都是连锁店，所以也具备物流网络的背景。这样的家庭餐厅也和便利店一样不依赖场所，通常开在带有停车场的人工地基上。

快递等小件配送网络的发展也催生了新的建筑类型。位于关越公路新座立交桥近旁的"53.物流立交枢纽"主要接收来自上信越方向的快递包裹，分流之后发往东京各区的分拣站和配送站。物流中心的一层装有大型的顶棚，建造了用于卸货的巨大平台。货物经传送带运往建筑物内部。建筑物上还建造了大型的行车坡道，楼顶就成了货车的停车场。根据用地地形的不同，建筑形式的布局多少会有些差别，但构成元素是固定不变的。位于首都高速公路沿线，邻近东京物流中心的"18.物流综合体"没有把坡道设置在建筑物四周，而是把坡道延伸进了建筑物的内部，同时还在卸货区的上方修建了员工宿舍。住在这栋员工宿舍里的人既住在大田区平和岛这个街区，同时又住在物流公司运输网络的一块虚拟用地上。如果去掉宿舍的部分，它就和"23.配送螺旋"十分相似。这些例子展示了物流公司在物流网络中发展出的建筑类型。

可以用"场所"概念解释的建筑历来都被认为是某个共同体的容器，但虚拟用地的出现消解了这项功能。这也就意味着我们需要重新考虑与共同体相关的诸多问题。新的建筑形态会催生新的共同体吗？或者，以各种网络为基础形成的社会性存在会吸收并完善原来的共同体吗？虽然现在还没有答案，但我们确信，建筑不能忽视后者的这种可能性。

shop design that is the important software for the management of the convenience store.
The existence of such convenience stores brings about the birth of a building type having no regard for the so-called concept of 'place.' They want to be able to melt into in any location, and so the issue of 'place' is avoided. This is horrifying from the viewpoint of architecture after Modernism, which so strongly believes in the concept of 'place' as the major explanation for design. If 'place' starts to lose its aura, then the concept of 'site' must also change. For each convenience store shop, there is not only the physical site, but it is also positioned within a site on a network. If we think of the physical site as 'real,' we can take the second type of site as 'virtual.' The way that this kind of virtual site makes a relation with architectural design is an issue needing full consideration. The buildings of Made in Tokyo include different types of networks with which to start this process.

Examples of virtual sites:
18. distribution complex
23. delivery spiral
24. bath tour building
53. dispersal terminal
58. family restaurant triplets

The coupling of architecture and the concept of 'place' previously acted as a container for 'community.' But the appearance of the 'virtual' site dissolves this couple, meaning that we have to work out what will happen to the issues inherent in that concept of community. Will those issues reappear as another guise of 'community,' or will they be directly propelled into a new idea of 'society'?

GUIDE BOOK

导览手册

功能：仓库+网球场

位置：千代田区饭田町

位于饭田町货运站内的一个纸品流通中心○二层设有用于接收纸品的铁轨，三到五层是用于仓储的仓库，一层是货车平台，货物从这里发出○160米 × 50米的楼顶上开设了网球学校，并配有直达电梯○场地位于东京市中心，离地30米，还能满足夜间比赛的需求

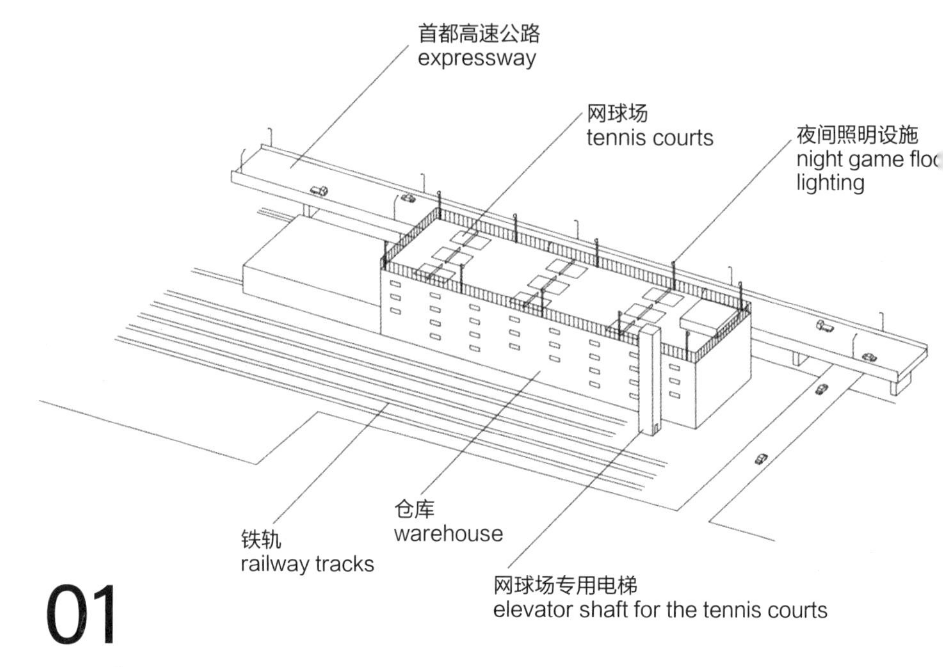

01

仓库球场

warehouse court

function: warehouse + tennis court
site: Iidamachi, Chiyoda-ku
- Iidamachi freight train station, paper dispatch centre
- large volume standing parallel to an expressway
- on the second floor is a platform for the delivery of paper from around the country; the third to fifth floors store the paper and the first floor is the truck yard for dispatch
- almost the entire floor area of the roof space, 160m × 50m is used as tennis courts
- the tennis courts can be accessed directly by an elevator shaft which stands independent to the main volume of the building
- there are lighting facilities for night games which also light up advertising; the tennis school is looking for students

功能：铁轨+店铺

位置：千代田区外神田

位于总武线秋叶原站西口前○高架铁路下的拱形结构也是商店街的拱廊○300米的高架下分出多家三层高的电器商店○经过拱廊街可以散步至神田一带

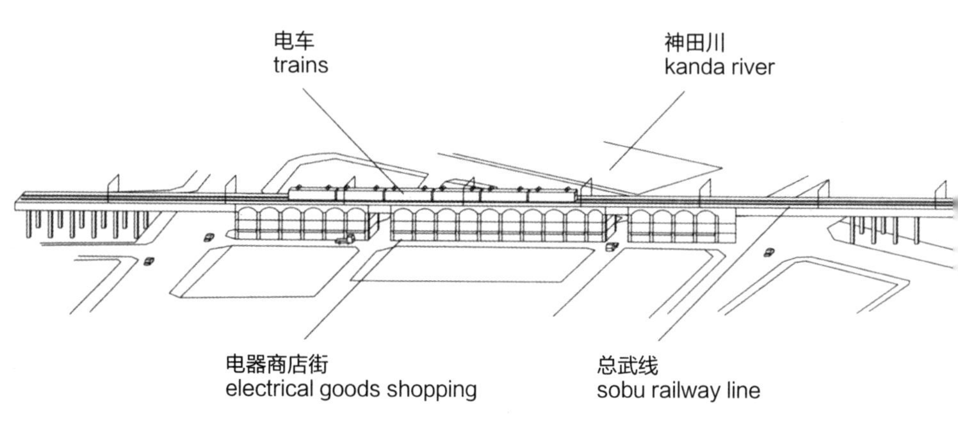

02

电器走廊

electric passage

function: railway bridge + shopping arcade
site: Sotokanda, Chiyoda-ku
- in front of the west exit, Akihabara Station, Sobu Line
- stacking and extension of the railway line and electrical goods district
- the railway tracks become a roof to 3 floors of electrical goods shopping
- 300m length of shopping arcade
- the scale of the frontage of each shop divides
this section of railway into smaller and smaller proportions

おかげさまで五十周年
大感謝祭
サトームセン
パソコン
オーディオフロア
3F
パソコン
プリンター
ワープロ
絶対安い!
2F
Sale
Sale
Sale
Sale
Sale
Sale
パソコン
ワープロ
サトームセン
GAME 1
電話機
大特価
NEC
パソコン

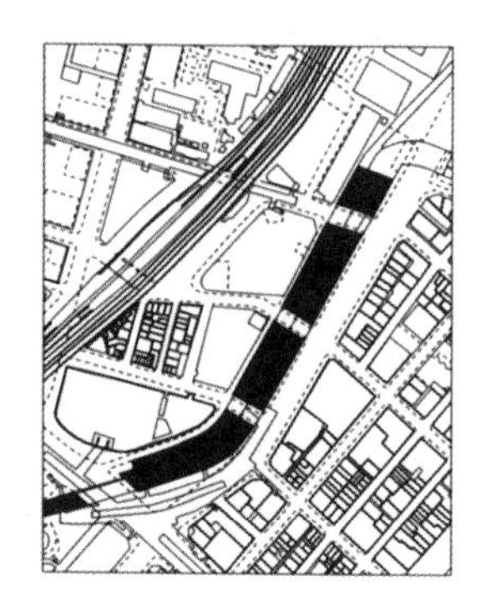

功能：高速公路+百货商店

位置：千代田区有乐町、中央区银座

在汐留川填河造陆的土地上建成〇蜿蜒的首都高速公路高架下开辟出约500米长的双层百货商店〇从高速公路上远眺银座夜景，蔚为壮观〇和八重洲地下停车场的坡道相连

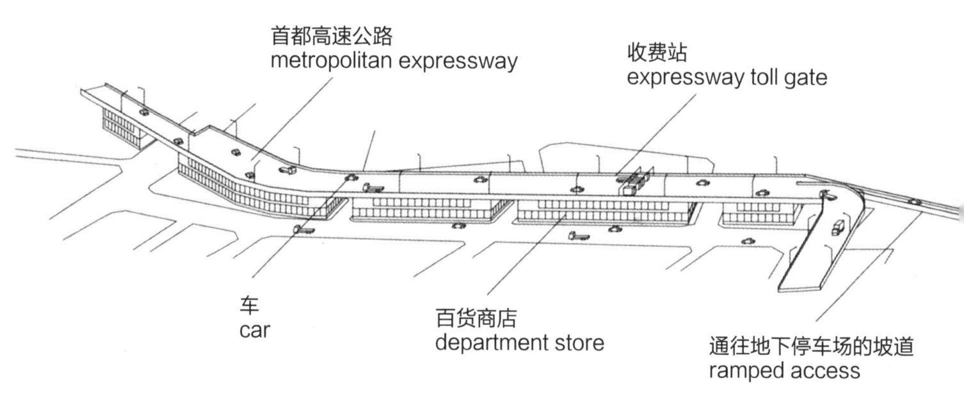

03

高速百货商店

highway department store

function: expressway + department store
site: Yurakucho, Chiyoda-ku and Ginza, Chuo-ku
- the department store spreads and extends together with the curving expressway
- the site fills the space of what was once Shiodome River
- the department store is two floors and extends for a length of 500m
- the expressway links to the line of lights from the underground parking facility
- due to the incoming traffic from the parking station,
the expressway is complicated by a toll gate

TOSHIBA
Asahi

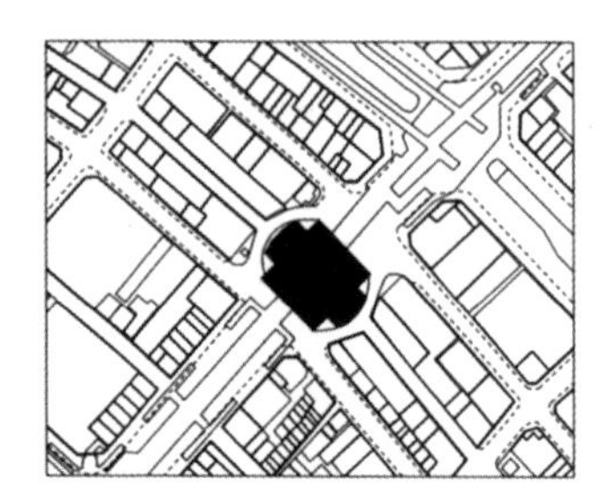

功能：地下通道+电影院+居酒屋+理发店+其他店铺

位置：中央区银座

位于晴海大道的歌舞伎座附近○在旧桥下方长50米的通道内开设了三家电影院和数家居酒屋○构造相似的店铺密布于通道两侧，标志着通道入口○由设计师土浦龟城设计

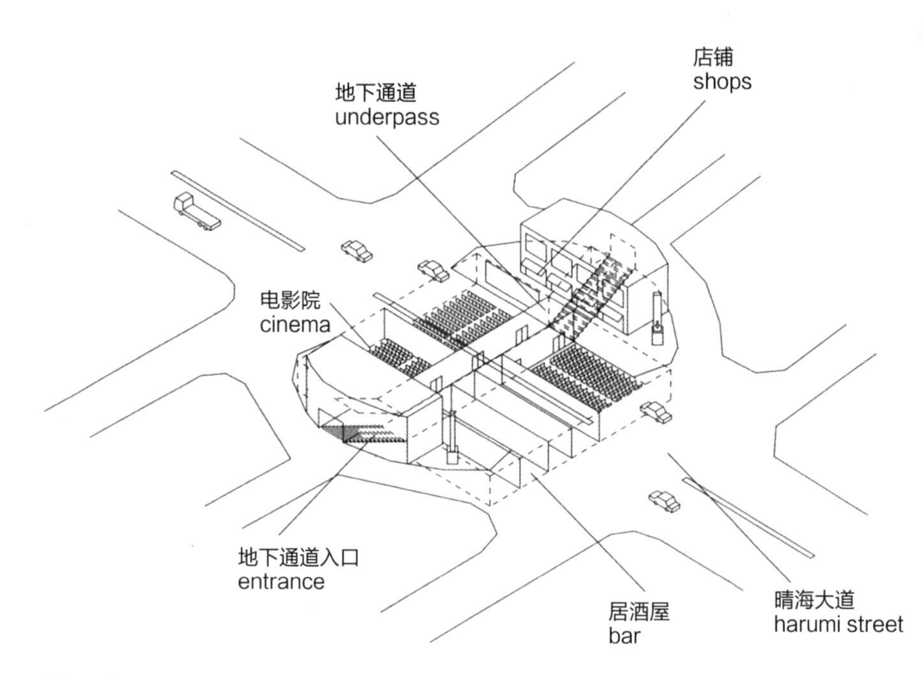

04

电影桥

cine-bridge

function: underpass + cinema + bar + barber + store
site: Ginza , Chuo-ku
- on Harumi street, near Kabuki-za theatre
- constructed under Harumi bridge at the time of the river being infilled, 3 cinemas and several sake bars align for 50m
- twin buildings face each other and sandwich the road to form underground entry points
- designed by Kameki Tsuchiura, pioneer Japanese modernist architect

シネパトス3
シネパトス2
シネパトス1
シネパトス
調髪料金
¥2,700
コールドパーマ ¥7000
アイロンパーマ ¥7000
食事処 三
¥620
SAPPORO

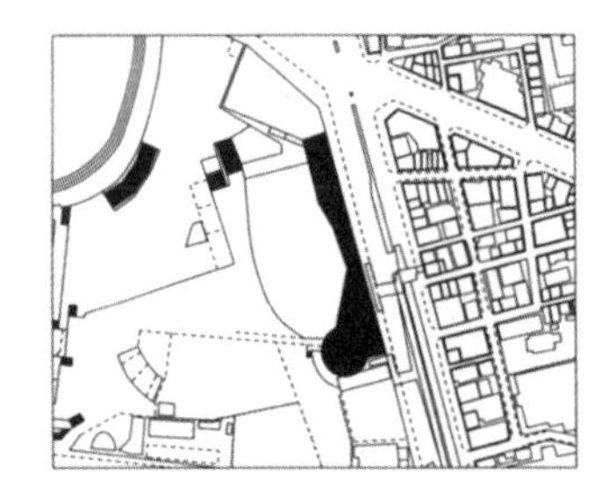

功能：过山车+游乐园大门+运动俱乐部+餐厅

位置：文京区后乐

后乐园游乐园的大门，入夜后就变身为照亮街道的照明装置○长约150米的店铺上方，过山车的轨道如巨龙一般蜿蜒盘旋○过山车和白山大道上的车辆并行○此外，还有转盘降落伞、跳楼机等可以感受不同速度和方向的娱乐设施

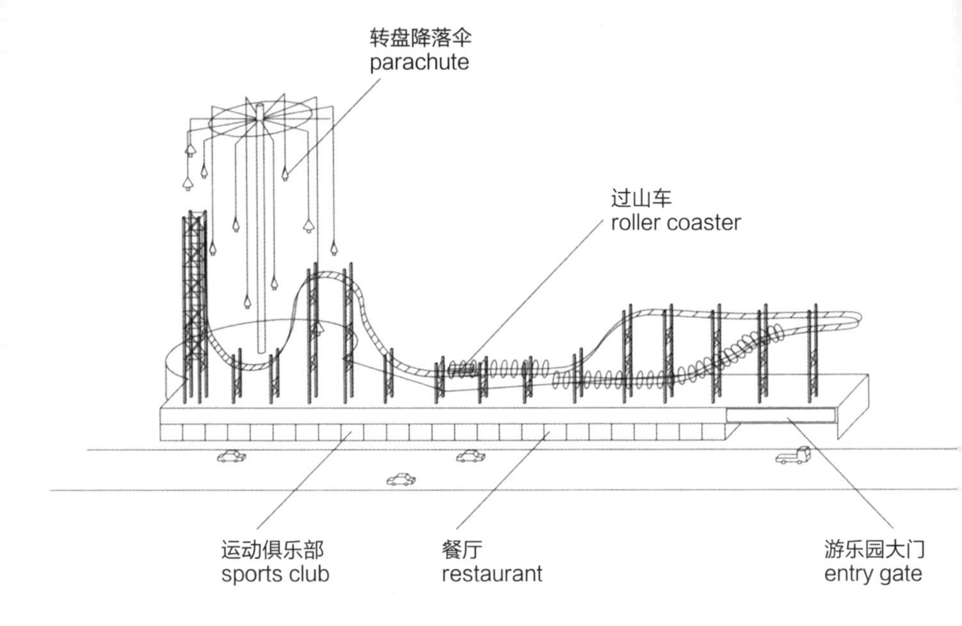

05

过山车楼

roller coaster building

function: roller coaster + entry gate+ sports club + restaurant
site: Koraku, Bunkyo-ku
- the building forms a gate to the Korakuen amusement park,
as well as being a lighting facility for the city at night
- above the 150m long strip of shops, the roller coaster railway slithers like a dragon
- the roller coaster railway aligns with Hakusan Street to race with the cars
- relativity can be experienced through the various other movements
also available in the area such as the 'parachute' and 'tower hakkaa'

功能：相机商店

位置：新宿区新宿

位于新宿东口站交通环岛前○覆盖建筑外立面的霓虹灯招牌上有设计各异的店铺名○招牌上不展示商品名称，只展示相机类型○隔壁的竞争店铺采用了相同的建筑样式○入夜后会变成车站前广场的街灯

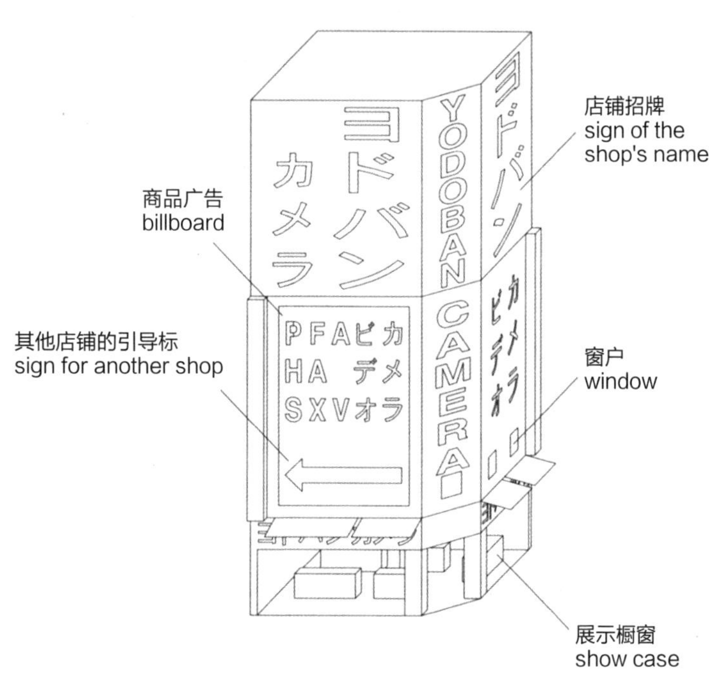

06

霓虹灯楼

neon building

function: camera store + billboard

site: Shinjuku, Shinjuku-ku

- in front of the Shinjuku East entry/exit interchange plaza
- neon signs wrap the external faces of the building, making various patterns by repeating shop names over and over again
- advertising lists types of product rather than brand names
- neighboring rival shares the same form
- in the evening, it becomes lighting for the open space of the city

さくらや
新宿東口店
MicroBoy
ヨドバシカメラ
YODOBASHI CAMERA
2F AVフロアー
1F テープ・携帯電話
B1 家電フロアー
Konica Plaza

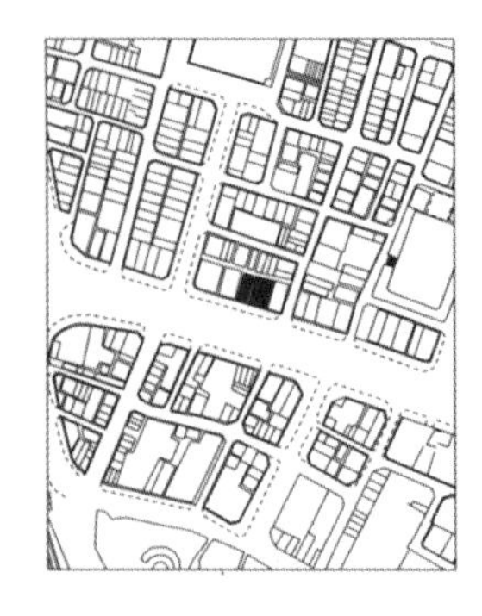

功能：柏青哥店+高利贷公司

位置：新宿区歌舞伎町

虽然是三座各自独立的建筑物，但并排的外观有如巴黎圣母院○教堂的外立面大都雕刻表现《圣经》的故事情节，而此处的外立面则被楼里商铺的招牌占据○两侧的铅笔楼，其租客大多是高利贷公司○在柏青哥店豪掷千金，借高利贷后再次入场，来客一进一出，建筑物彼此对称，构成了独特的城市生态系统

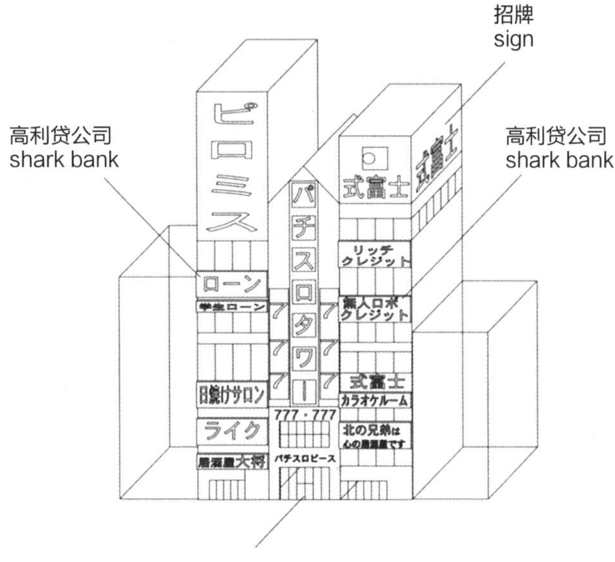

07

柏青哥教堂

pachinko cathedral

function: pachinko parlour + shark bank

site: Kabuki-cha, Shinjuku

- they are 3 separate buildings, but if viewed as a unit, they take on the appearance of Paris' Notre Dame cathedral
- instead of ornament showing the story of the Bible, this complex pulses with information banners advertising the internal activities
- the tenancies of the side tower buildings are almost completely made up of shark banks loaning money at extremely high interest rates
- one ecological system of the metropolis is formulated as a symmetrical package; an endless cycle of losing money at pachinko, loaning money, losing money...

¥en shop
武富士

功能：风俗店

位置：新宿区歌舞伎町

歌舞伎町一栋多租户的大楼○租户多为风俗业从业者○整座建筑物的外立面布满了店铺的招牌○附近还有不少同类建筑物

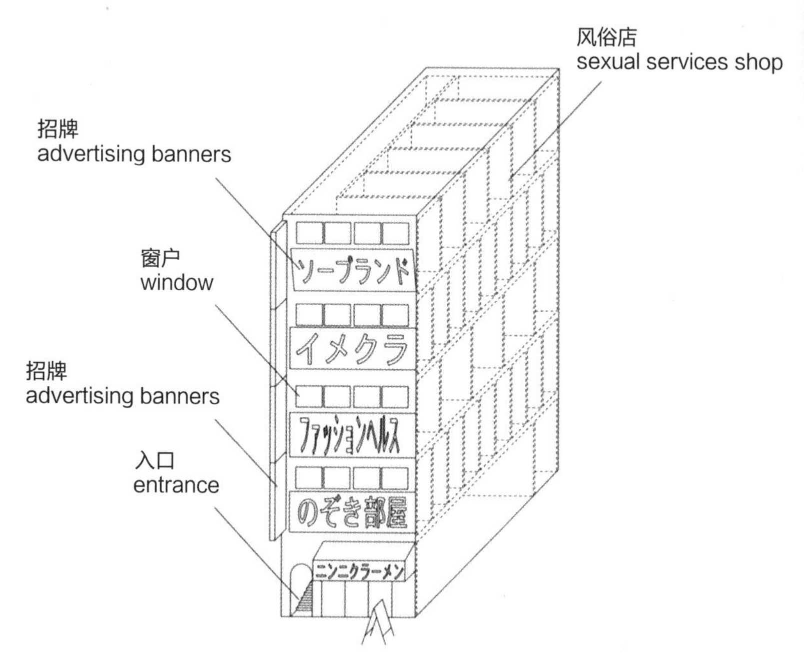

08

风俗楼

sex building

function: sexual services

site: Kabukicho, Shinjuku-ku

- mixed tenancy building in the notorious Kabukicho
- each floor contains a series of rooms making up the sexual services shop
- each level of the front facade is plastered by a lit up banner advertising the shop
- this building type is typical in the area

新宿ノ浮気現場
リッチドール
3F
参千円
2F
ラーメン
アダルト
ビデオ
美乳の女神
リッチドール
30分
6500
びんびん
天才少女大集合!
2F
新宿
B2
3F
ヘルス界の殿堂
新宿 リッチドール

功能：卡拉OK包厢

位置：港区六本木

集合式卡拉OK包厢的变奏○集合式卡拉OK包厢还有塔式（涩谷区宇田川町）、公寓式（杉并区方南）等形式○在大堂柜台登记之后，客人会被引导到单独的卡拉OK包厢○从挂着红窗帘的窗户向外望去，六本木的街景十分迷人

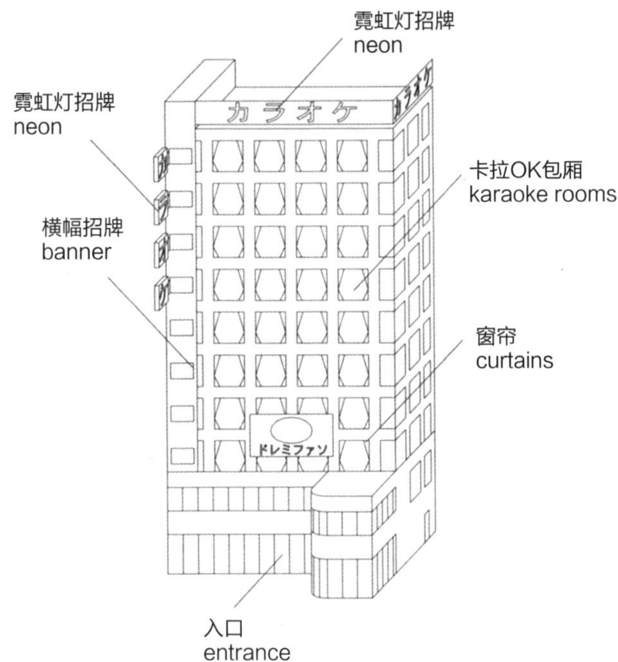

09

卡拉 OK 宾馆

karaoke hotel

function: collective karaoke rooms
site: Roppongi, Minato-ku
- an urban variation of the karaoke type
where the individual rooms are collected like a hotel
- other urban types include the tower type (Udagawacho, Shibuya-ku)
and apartment type (Honan, Suginami-ku)
- after checking in at the lobby counter, karaoke in the private rooms
- view the cityscape of Roppongi through the luscious red curtains

Club

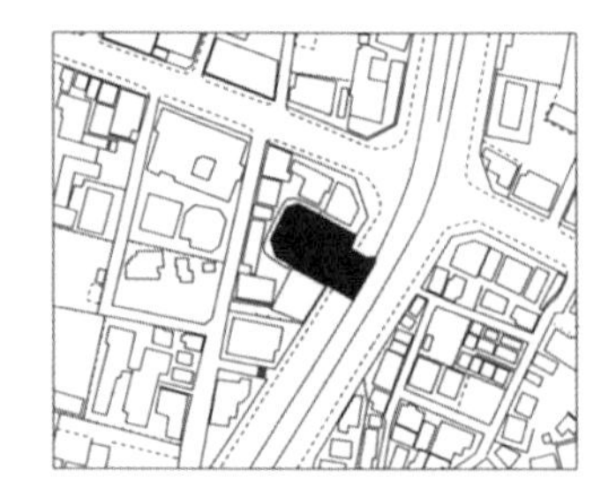

功能：办公室+员工宿舍+车库

位置：港区六本木

唯一与首都高速公路直接相连的建筑形态○普通公路和首都高速公路通过建筑物内部的斜坡互通○连接首都高速公路的桥兼有巡逻车停车场的功能○车库上方的员工宿舍属于道路公团的大楼○护国寺出口附近也有类似的建筑

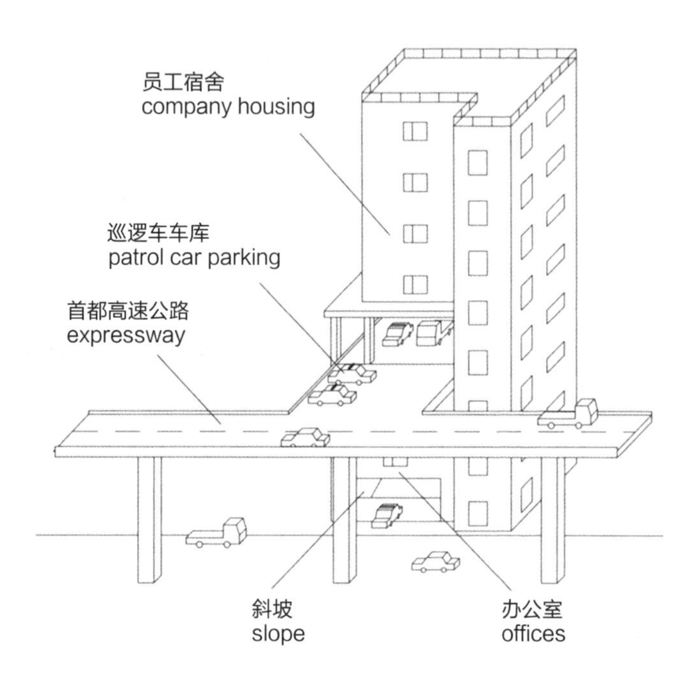

10

首都高速公路巡逻站

expressway patrol building

function: office + company housing + patrol car parking
site: Roppongi, Minato-ku
- the only kind of building whose vehicles could
possibly access directly to the expressway
- a ramp inside the building connects the expressway with the normal road below
- the bridge between building and expressway acts also as parking for the patrol cars
- company housing on the upper levels of the building
belonging to the metropolitan expressway
- another similar example near the Gokokuji expressway interchange

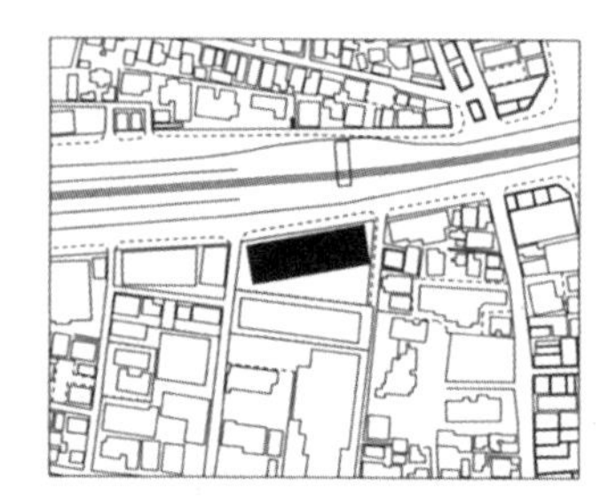

功能：大使馆

位置：港区西麻布

位于六本木大道沿线○大楼里入驻了十几个国家的大使馆，规模巨大，出入口却很少○入口旁边飘扬的各国国旗是大使馆的唯一标志○此处也是东京高昂地价的象征

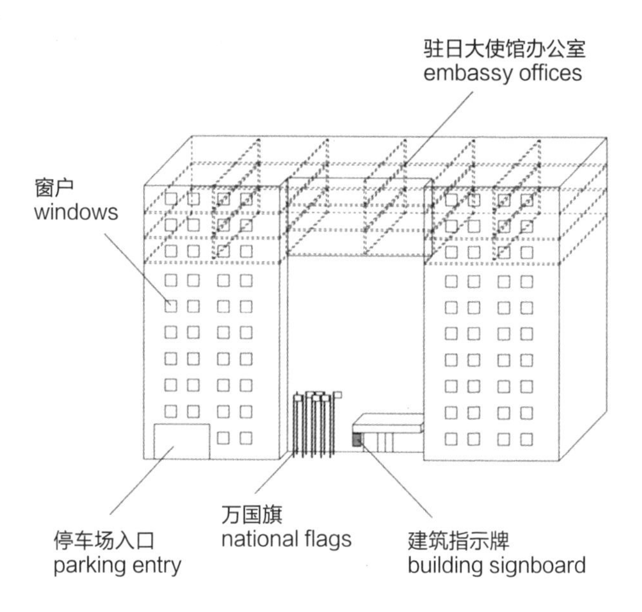

11

大使馆楼

embassies building

function: embassies

site: Nishiazabu, Minato-ku

- on Roppongi Street
- nineteen countries share three floors of a massive building with small windows
- the only embassadorial display to the street is the national flags at the entry
- symbolises the high price of land in Tokyo

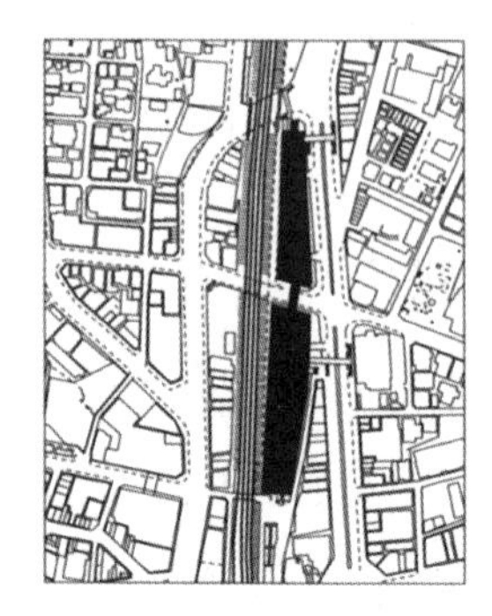

功能：公园+停车场

位置：涩谷区涩谷、明治神宫前

位于明治大道与山手线和埼京线之间○总长约330米，宽约20到30米的细长人造空间○上层是公园，下层是停车场。若干车道横穿整个空间○不知为何，经常能在这里看到抱在一起的中年男女，难道是个偷情圣地?

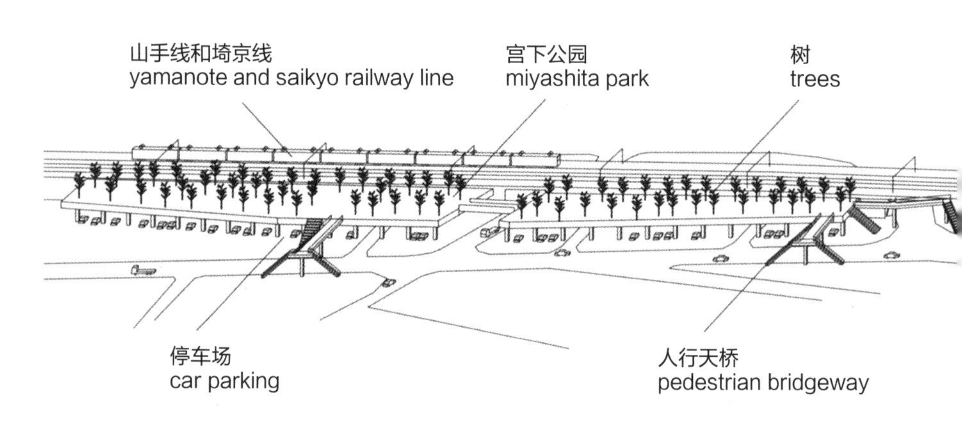

12

公园停车场

park on park

function: public park + car park
site: Shibuya Jingumae, Shibuya-ku
- a narrow artificial ground standing between Meiji street and the railway tracks of Yamanote line and Saikyo line
- the park width is between 20 and 30 metres, the length is 330 metres
- Miyashita park forms the roof level of the carpark
- the park is accessed from pedestrian bridgeways over Meiji street
- the park continues over the street which divides the carpark below

功能：公交车车库+集合住宅
位置：涩谷区东部
位于涩谷站附近，东横线和涩谷川边上○都营住宅呈板状，其底部架空层是都营公交车车库○一个结构单位可容纳两辆公交车和两户住户○一辆公交车的大小等于一个住户单元的大小○公交车和人以同样的形式被“收纳”到同一座建筑物里

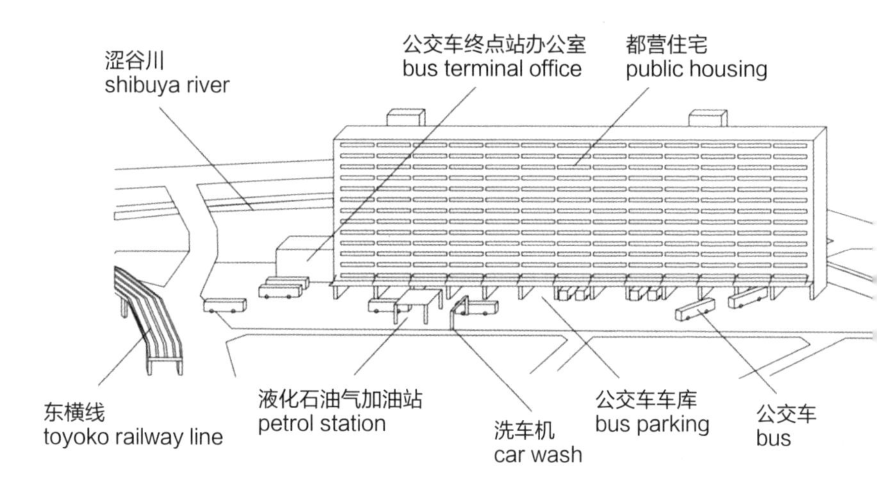

13
公交住宅区
bus housing

function: bus terminal + apartment housing
site: Higashi, Shibuya-ku
- near Shibuya Station, visible from the raised tracks of the Toyoko railway line
- on top of a metropolitan bus terminal stands a slab volume of public housing
- each structural bay width allows for 2 buses and 2 apartments
- 1 bus length = 1 apartment length
- both buses and people are put into the same form

功能：高尔夫球练习场+办公室+车库

位置：目黑区目黑

位于目黑川沿岸，区民活动中心对面○高尔夫球练习场横跨出租车公司的办公室和停车场○练习高尔夫的人们从办公楼的楼顶朝目黑川的方向挥杆○球会被预先架设在停车场上方的网拦住，滚回练习场○到了晚上，高尔夫球练习场变身为一个发光的绿色笼子，在河面上留下倒影

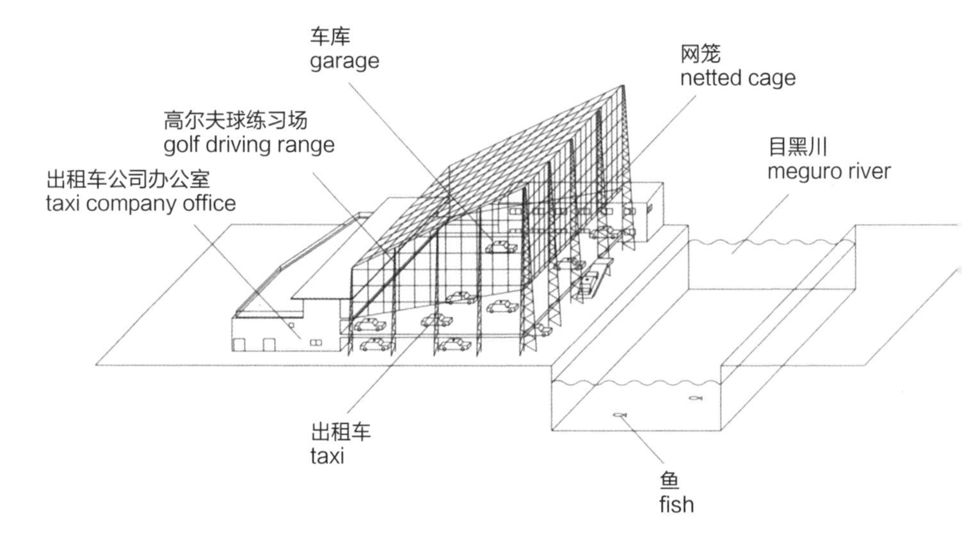

14

高尔夫出租车站

golf taxi building

function: golf driving range + taxi office + taxi parking garage

site: Meguro, Meguro-ku

- along the Meguro river, facing the Meguro Community Center
- from on top of the taxi company office, the golfers practice driving towards the river
- the ceiling of the taxi parking is a huge sloping, netted cage, through which driven balls fly and roll back towards the office and golfers
- in the evening, it becomes a green cage of light, reflected in the water surface of the river

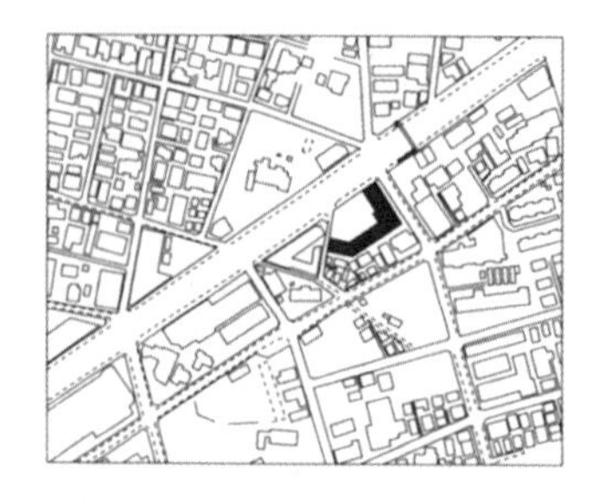

功能：混凝土工厂+员工宿舍

位置：目黑区碑文谷

位于目黑大道大荣超市碑文谷店附近○闪着银色亮光的混凝土设备和司机宿舍相连○下方是随时准备为东京输送混凝土的搅拌车○把人和机器的空间有机结合，打造出一座奇妙的工作生活两用建筑

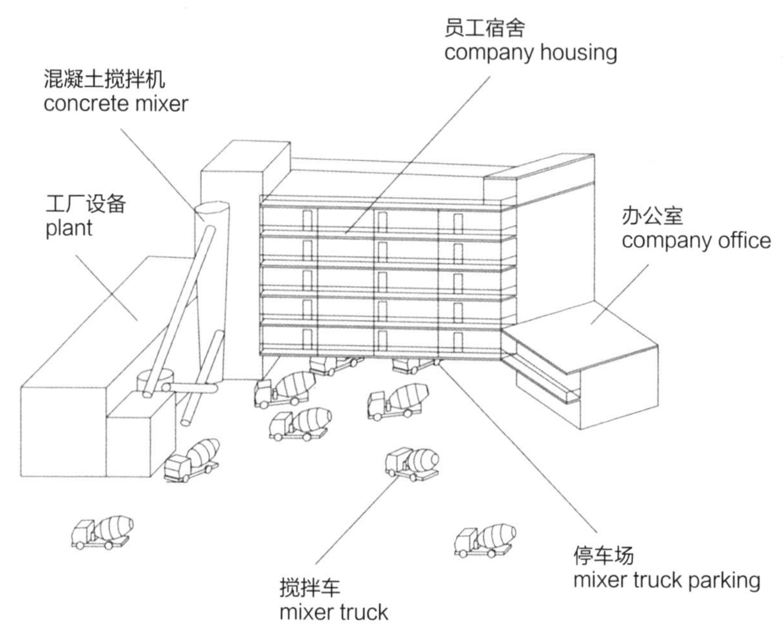

15

混凝土公寓

nama-con apartment house

function: concrete batch plant + company housing
site: Himon-ya, Meguro-ku
- near Daiei supermarket and aligning with Meguro street
- the docking together of the silver shine of the concrete plant
and the mixer truck drivers housing
- the mixers wait at the bottom of the plant to service the city
- the packaging together of the workplace
and home makes an veritable man-machine system

olivetti

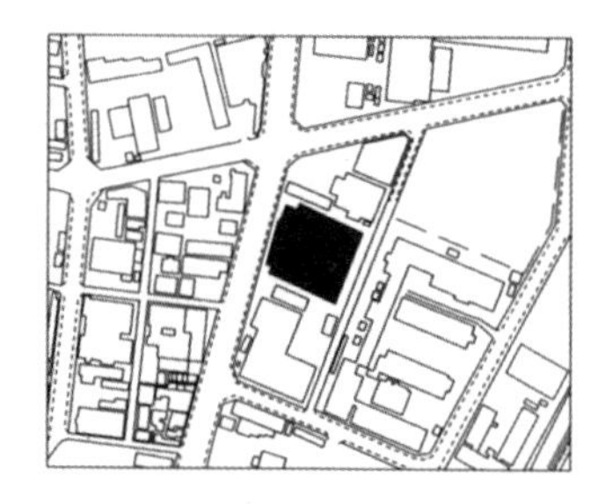

功能：展示厅+办公室+修理厂+车库

位置：品川区东大井

位于湾岸大道沿线，鲛洲驾考考场附近○建筑物底层是展示厅和修理厂，上方是巨大的车库，环绕在四周的螺旋状车道不断向上延伸○建筑总面积中，人与车的比例发生了逆转○车道盘旋而上，突然出现在城市的高空

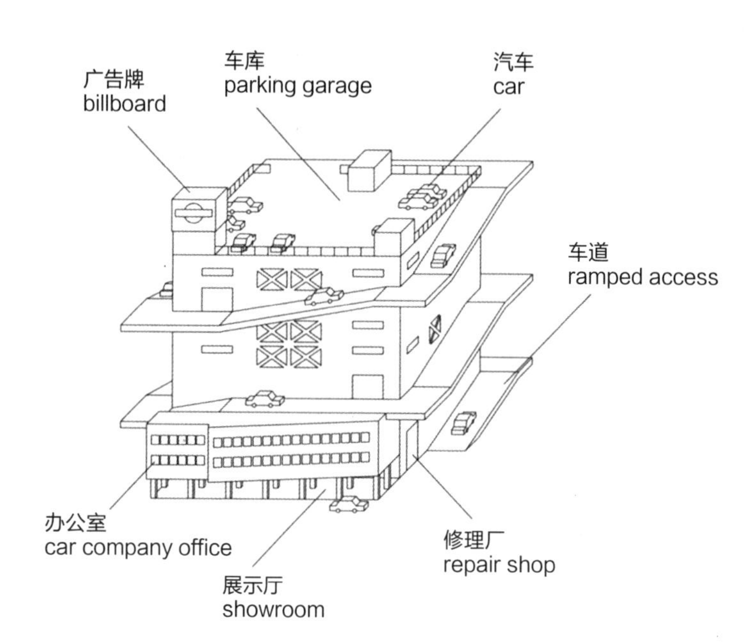

16

汽车塔

car tower

function: car showroom + office + repair shop + parking garage
site: Higashi-oi, Shinagawa-ku
- along Wangan street and near the Samezu car licencing board
- single package-building for all aspects of car service
- an external car access road spirals around
the tower made up of showroom, offices and carpark
- a proportional flip between a building's usual numbers
of people compared to numbers of cars
- suddenly, a road flies through the middle of the city's air space

日産プリンス
東京
NISSAN PRINCE TOKYO

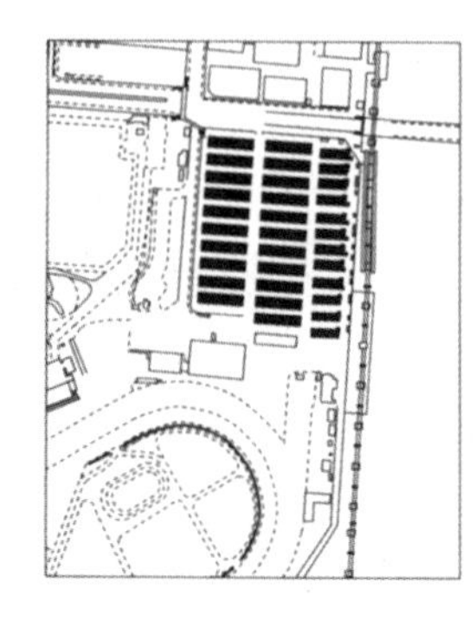

功能：马厩+驯马师宿舍

位置：品川区胜岛

毗邻大井赛马场，可以从东京单轨电车上看到○33间样式相同的厩舍整齐地平行排列在一起○一层为赛马的马厩，二、三层为驯马师宿舍○人和马住在一起

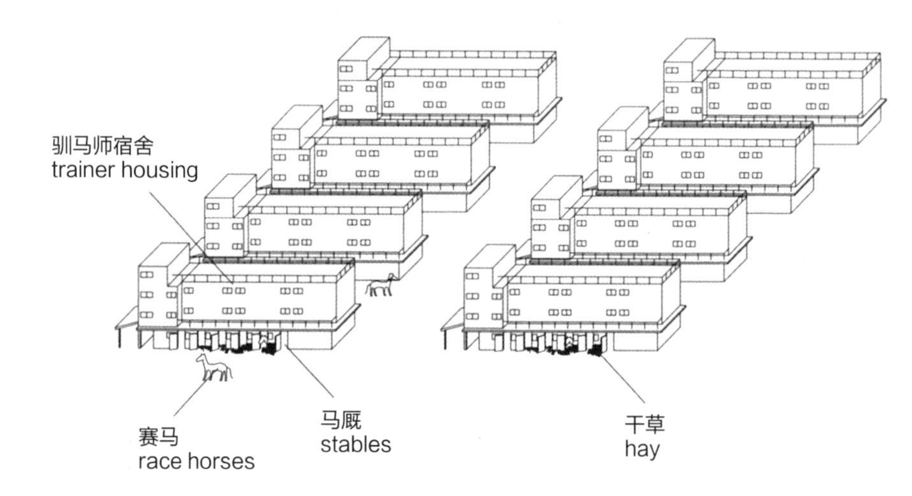

17

马公寓

horse apartment house

function: stables + trainer housing
site: Katsushima, Shinagawa-ku
- next to the Oi horse racing track
- visible from the monorail
- one building is multiplied 33 times and neatly arranged in parallel
- race horse stables are on the ground floor, trainer housing is above
- horses and people live together

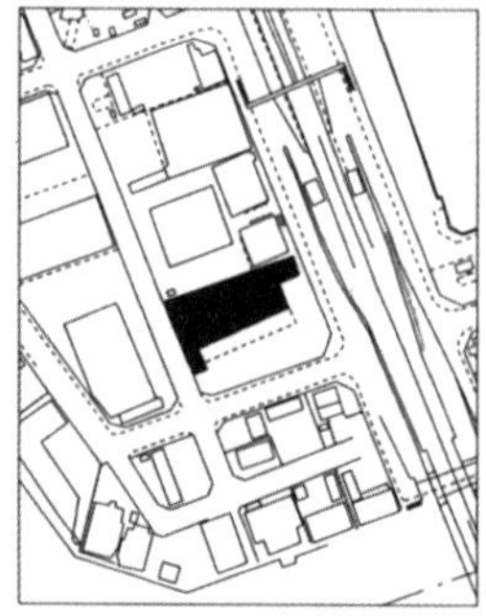

功能：办公室+车库+仓库+员工宿舍

位置：大田区平和岛

位于首都高速公路沿线，东京物流中心附近○物流公司的配送站上方是员工宿舍○斜坡车道一直延伸到建筑物内部

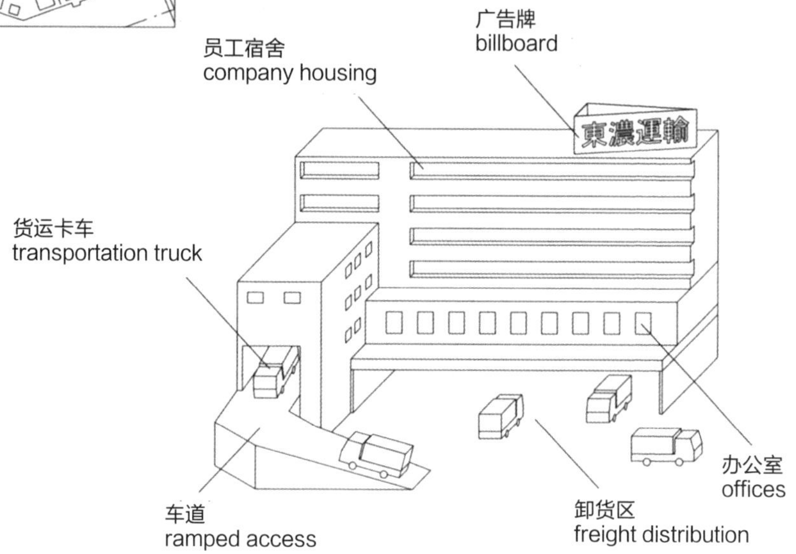

18

物流综合体

distribution complex

function: office + parking + distribution pool + company housing
site: Heiwajima, Ota-ku
- aligning the expressway, part of an area in the city of delivery interchange points
- this distribution company centre includes employee housing above
- the roadway ramp eats its way into the building

SEINO
カンガルー宅配便

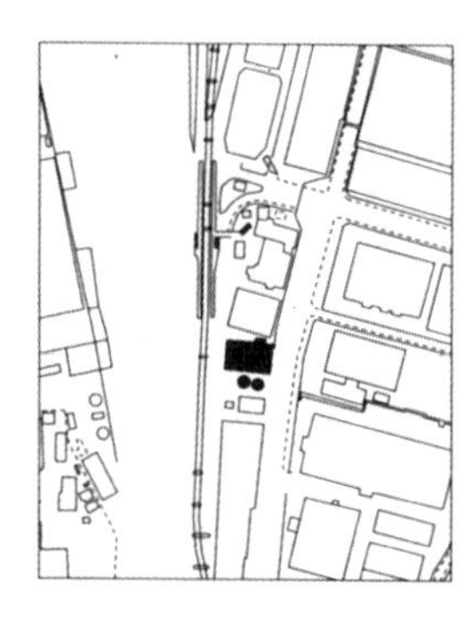

功能：地区冷暖气设备

位置：大田区羽田机场

位于东京单轨电车修理厂站附近○为航站楼、飞机修理厂等周边的机场相关设施提供冷暖气的地区冷暖气供应设施○内部装配冷热交换器，建筑物体积巨大○屋顶有完全外露的大型冷却塔○外墙上架设管道，与其他设施相连

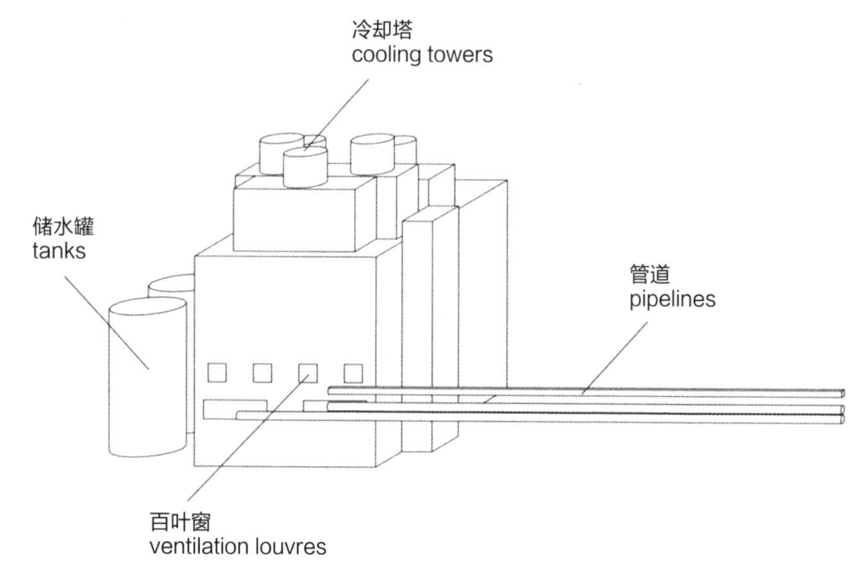

19

空调楼

air-con building

function: district cooling and heating plant

site: Haneda-kuko, Ota-ku

- near the monorail stop called 'Haneda Airport Maintenance Area Station'
- facility for heating and cooling the surrounding buildings related to the airport, such as aeroplane maintenance factories
- a huge volume made up of mechanical equipment for cooling and heating exchange
- large, open cooling towers on the roof terrace
- pipelines extend from the building to connect into surrounding buildings

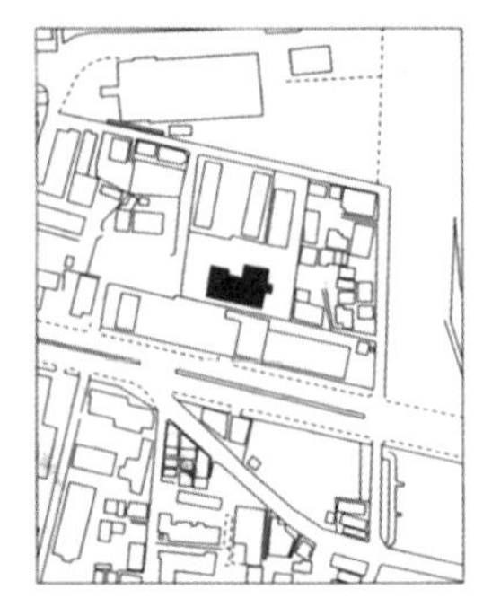

功能：广告牌+集合住宅

位置：大田区羽田旭町

位于羽田机场附近○广告牌面积巨大，飞机上的乘客从空中也能辨认，广告牌的底座是一栋高层公寓○广告牌的大小几乎和下方的公寓差不多○入夜后，发光的广告牌格外显眼○还有类似的以办公楼为底座放置广告牌的例子

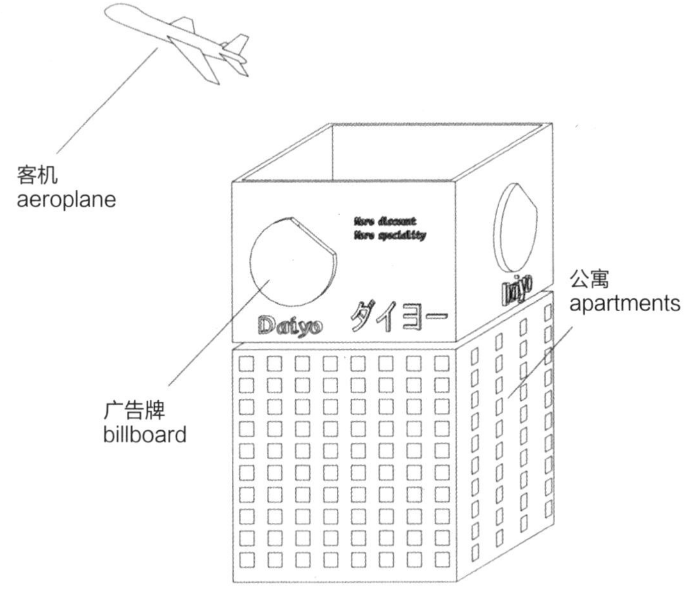

20

广告公寓

billboard apartment house

function: billboard + apartment house
site: Haneda-asahicho, Ota-ku
- stands near Haneda airport
- the scale of the huge billboard suits the view from the aeroplane; the apartments' role is to hold up this billboard
- the billboard and the building below are almost the same size
- the billboard is almost half again the size of the building below
- in the evening, the flashing neon sign floats above the rest of the city
- sometimes, this kind of billboard is held up by office buildings

Daiei
More discount
More "specialty"
ダイエー

功能：神社+商店

位置：大田区田园调布

位于东横线和目蒲线多摩川园站附近○神社的一部分建在人工地基上○人工地基下方是出租的店铺，上方则是神社院落和停车场○类似的例子是把神社建在餐饮店上方（台东区根岸）

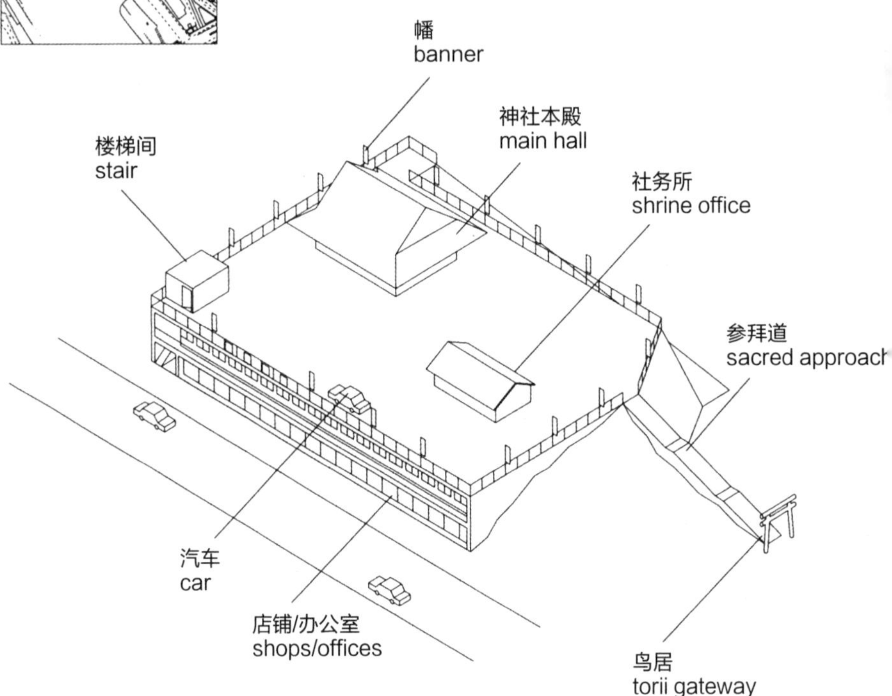

21

神社楼

shrine building

function: shrine + shops/offices

site: Denen-chofu, Ota-ku

- near Tamagawaen station, Toyoko and Mekama lines
- the shrine precinct is an artificial ground on a rooftop
- the lower levels contain rented commercial space,
on the rooftop is the shrine, and carparking
- a similar example of shrine over a restaurant building is in Negishi, Tailo-ku

神 社
養老乃瀧
Yoro No Taki
厄除、

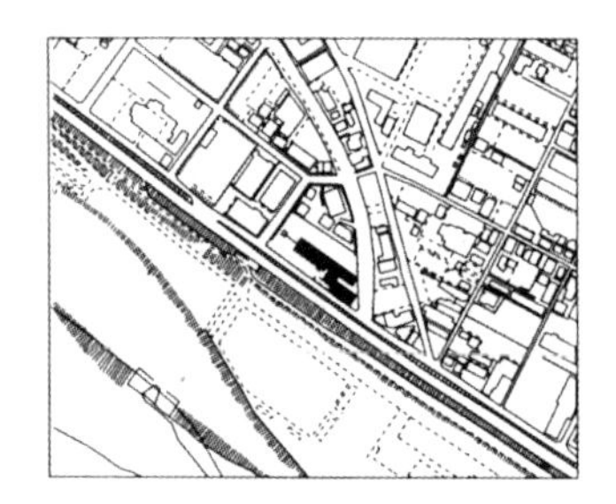

功能：办公室+员工宿舍+建材堆积场+车库

位置：世田谷区玉堤

位于多摩堤大道沿岸○架空层上方是土木工程公司的办公室和员工宿舍○整个底层则用作建材（土）堆积场和车库

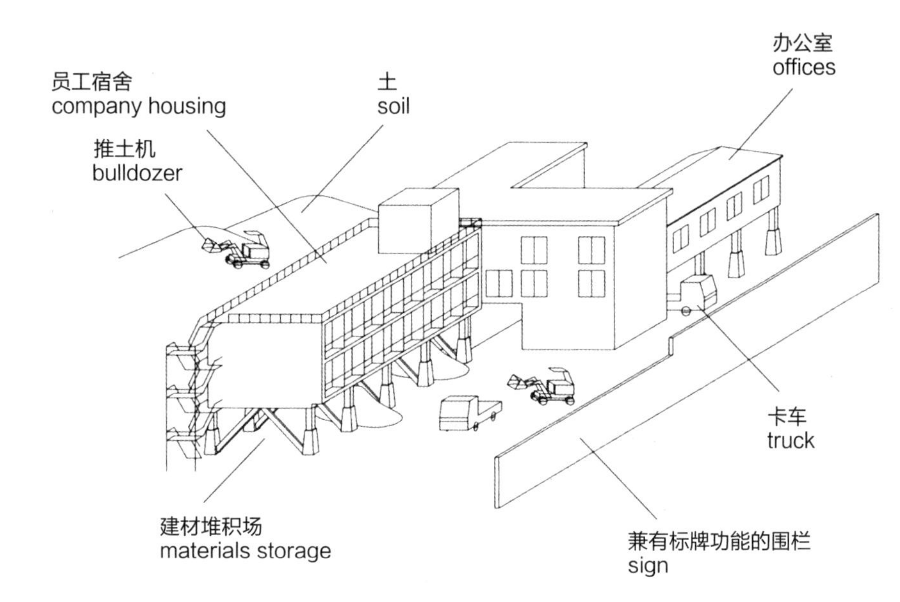

22

渣土公寓

sand apartment house

function: office + company housing
 + building materials storage + truck parking
site: Tamazutsumi, Setagaya-ku
- stands alongside the Tamazutsumi road, and the wide bank of the Tama River
- company housing and office of the earthworks
construction company floats above, on piloti
- the full extent of the ground level of the site is free to be used
for building materials and soil storage, truck parking and so on

功能：办公室+车库+仓库

位置：杉井区和泉

位于井之头线附近○快递的配送中心○蜿蜒的斜坡车道往返进出建筑物○乍一看像极了勒·柯布西耶设计的卡彭特视觉艺术中心

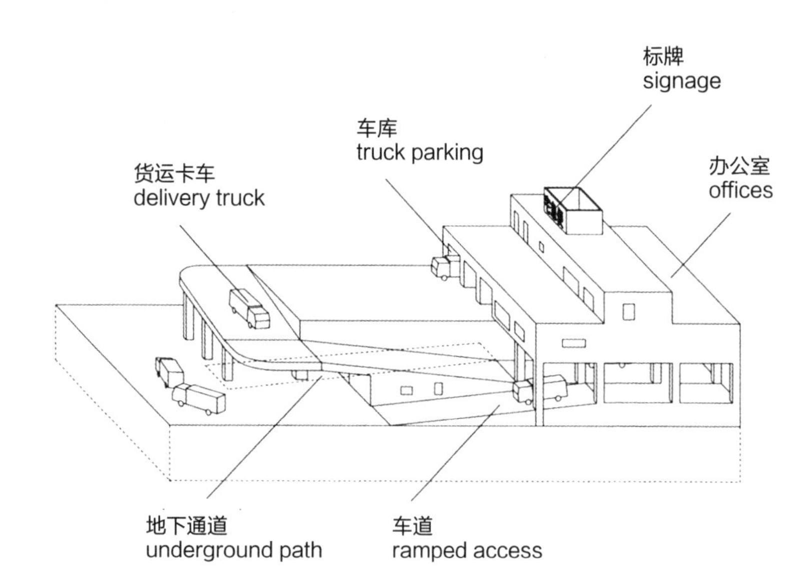

23

配送螺旋

delivery spiral

function: office + truck parking + courier storage centre
site: Izumi, Suginami-ku

- alongside Inokashira railway line
- delivery centre for courier mail
- vehicular ramps fold in and out, through the volume of the building
- it reminds us of Le Corbusier's Carpenter Center

功能：公共澡堂+桑拿房+自助洗衣店+便利店+民宅

位置：杉并区高圆寺南

位于环状7号线沿线○面朝公路的一侧的一层开设有便利店和自助洗衣店，二层是公共澡堂，三层是桑拿房○入浴、洗衣、购物等需求在同一座建筑物里得到满足○单身人士夜晚的社交场所

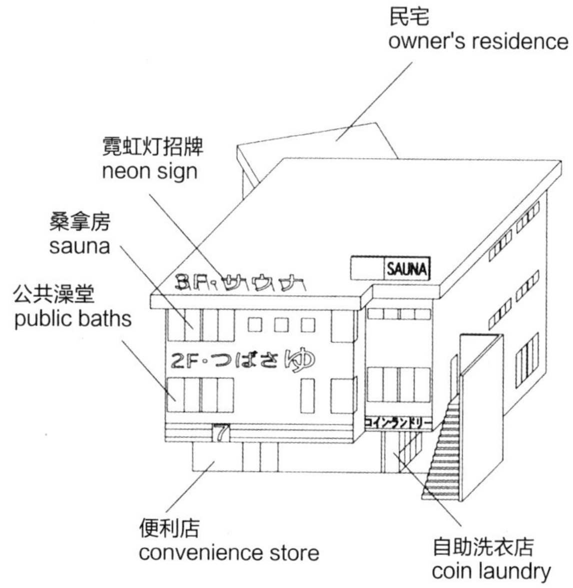

24

澡堂旅游楼

bath tour building

function: public bath + sauna + coin laundry
+ convenience store + owner's residence
site: Koenji-minami, Suginami-ku
- directly faces the busy Kannana (seventh) traffic ring road
- on the first floor linking to the street is a 24 hour convenience store and coin laundry, on the second floor is the public bath facility and on the third floor is the sauna
- a package of activity, comprising a sequence of bathing, washing, shopping
- a singles' night spot

SAUNA
サンデッキ
2F・つがさゆ
ELEVEn
コミュニティー・ランドリー

功能：修理厂+办公室+车库+休息室

位置：武藏野市吉祥寺南町

位于井之头大道沿线的出租车公司○曾是日产公司的修理厂○一层是修理厂，二层是办公室和车库，三层是司机们的休息室○人车同床，各自安眠

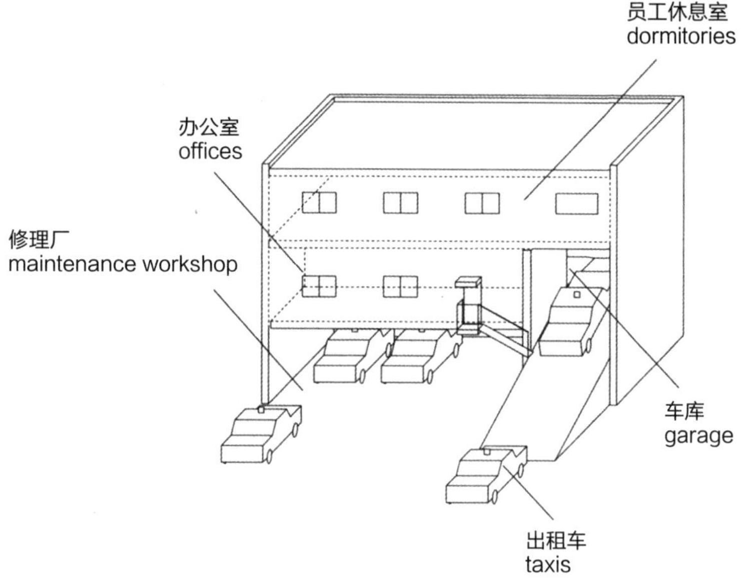

25

出租车楼

taxi building

function: maintenance workshop + office + garage + dormitory

site: Kichijoji-minamicho, Musashino-shi

- the taxi company building stands alongside Inokashira Street
- the building used to be the Nissan repair factory
- on the first floor is the maintenance workshop, on the second floor is the taxi office and parking, on the third floor is the temporary sleeping area for the drivers
- humans and cars are sleeping next to each other

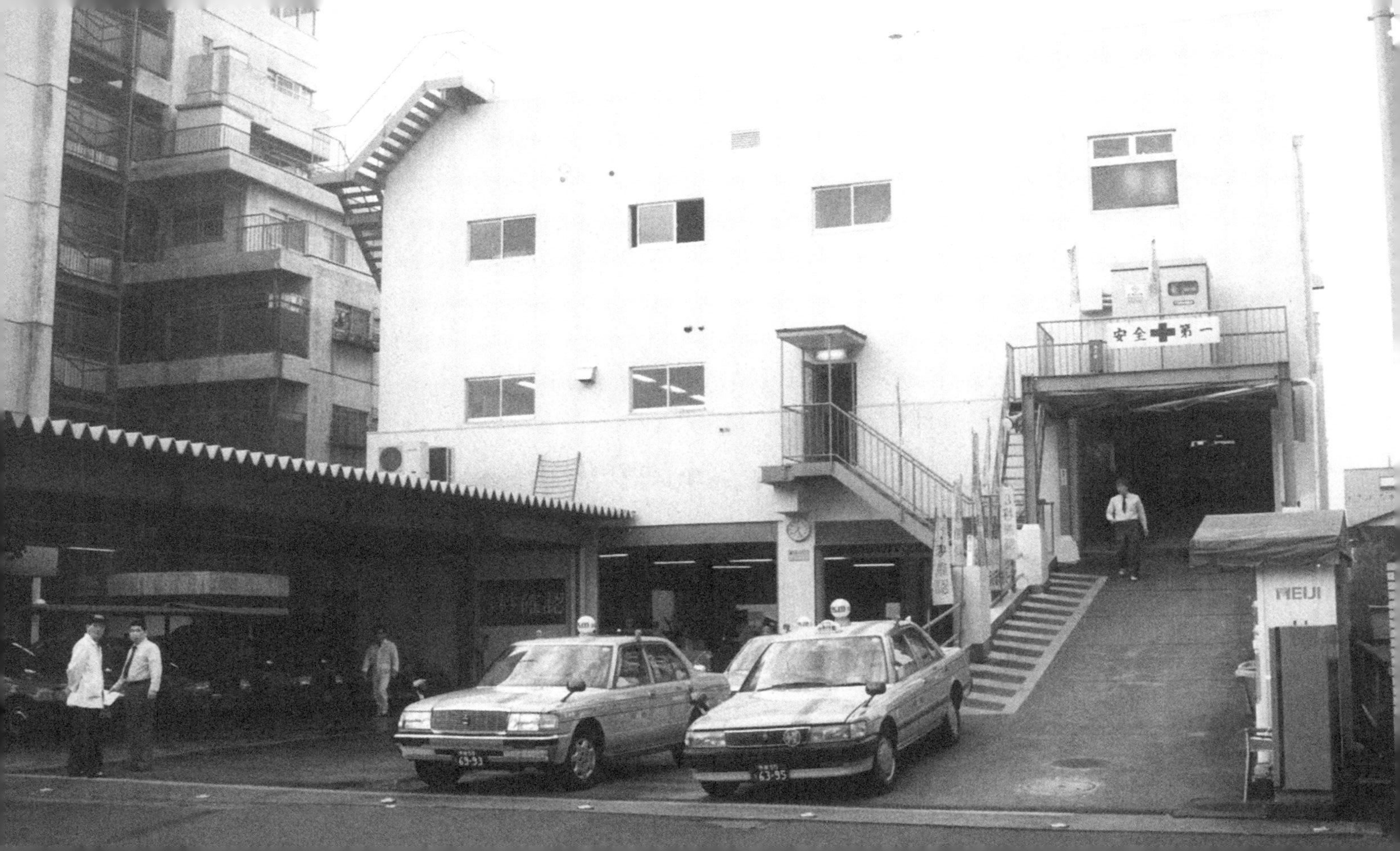
安全第一
MEIJI

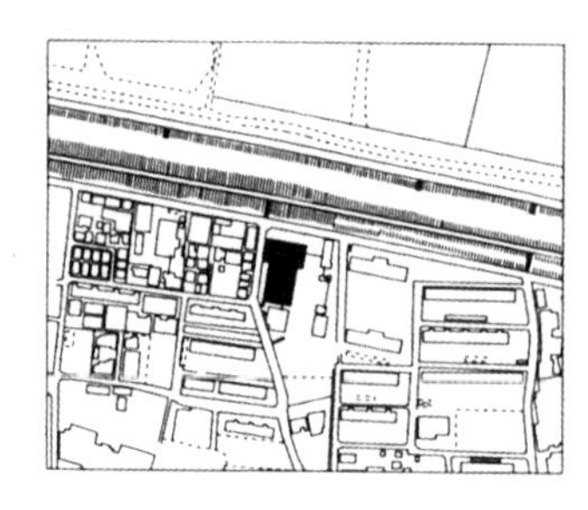

功能：修理厂+办公室+车库

位置：板桥区舟渡

位于荒川防波堤沿岸○一层是修理厂，二、三层设有办公室，二到六层都是车库○因为是货车专用的停车塔，所以层高要比普通汽车塔高出一些

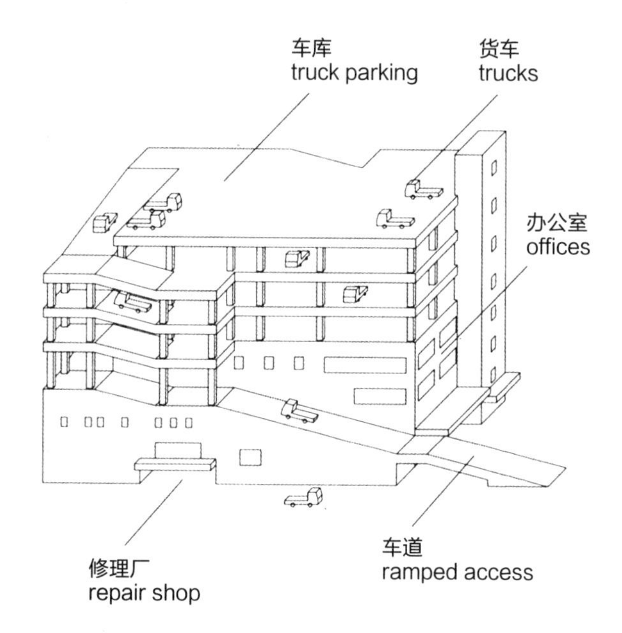

26

货车塔

truck tower

function: repair shop + office + truck parking
site: Funado, Itabashi-ku
- stands alongside the Arakawa River embankment
- on the first floor is the repair shop, on the second and third floors are offices and from the second to the sixth floors is truck parking
- because the parking is for trucks, the floor to ceiling heights are larger than for a ‘car tower’

城西ふそう

功能：高速公路立交桥+网球场

位置：足立区西加平

首都高速公路加平匝道的两个螺旋圆环之一，与环状7号线相连○由车道围成的中庭被用作网球场○另一个圆环的中庭则被改造成道路公团巡逻车的停车场○外侧用仿石材料做出粗面石工效果○巴洛克风格的中庭被钢柱回廊包围

路灯
street light

巡逻车停车场
patrol car parking

汽车
car

首都高速公路
expressway

网球场
tennis courts

车道
ramped access

27

立交球场

interchange court

function: repair shop + office + truck parking
site: Nishikahei, Adachi-ku
- one of the interchange spirals at Kahei,
which links the expressway with the Kannana (seventh) traffic ring road
- the courtyard enclosed by the spiralling ramp has become tennis courts
- the courtyard on the other side is car parking for the expressway patrol cars
- the ramp is enclosed below with an external fake stone block facade,
in an rustic style
- an iron colonnade surrounds the courtyard, in an baroque style

功能：加油站（一层、二层）+办公室
位置：新宿区西新宿
位于新宿公园塔附近的一个立交口○位于建筑一层和二层的加油站分别和立交桥上下两层的车道相连○虽然同属一家石油公司，但两家加油站的营业时间却不同

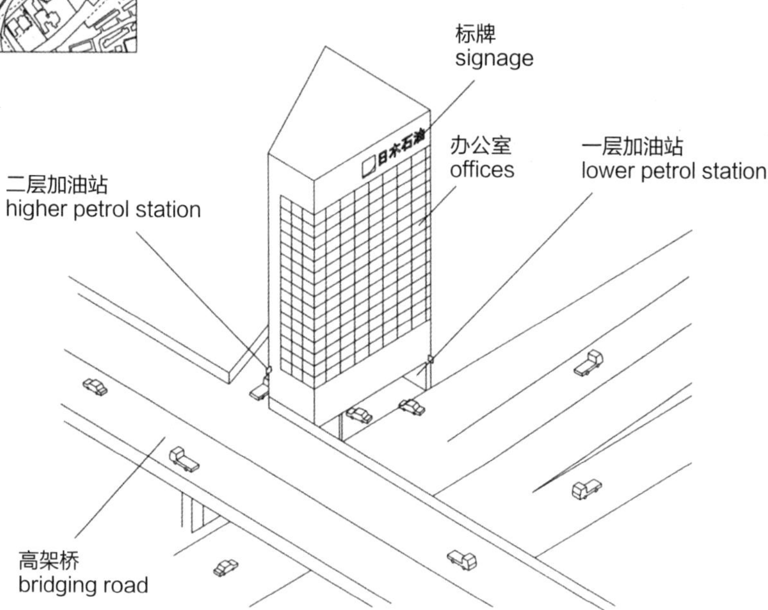

28

跃层式加油站

double layer petrol station

function: petrol station + office
site: Nishishinjuku, Shinjuku-ku
- at a split intersection near Park Tower Hotel
- double layer petrol stations can be accessed from both the lower and higher roadways
- although they are run by the same petroleum company, the 2 stations have differing operating times

NiSSEKI

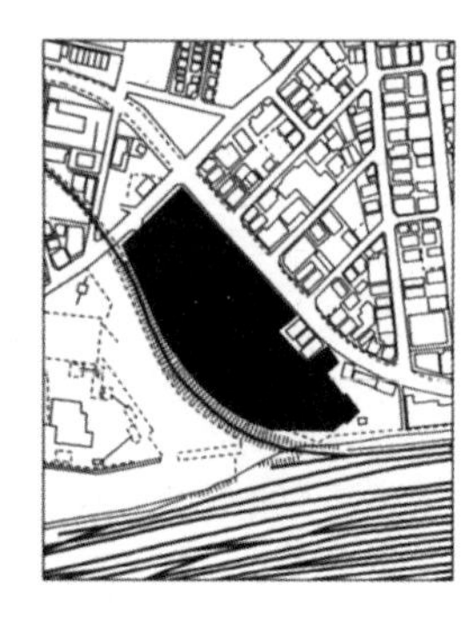

功能：超市+驾校

位置：葛饰区金町

双层超市楼顶开设了驾校○驾校使用了一部分尚未完成收购的建筑用地○这块建筑用地被弯曲的铁路包围，其形状直接体现在驾校的设计中○用于练习坡道起步的斜坡下方就是和外部道路相连的斜坡出口

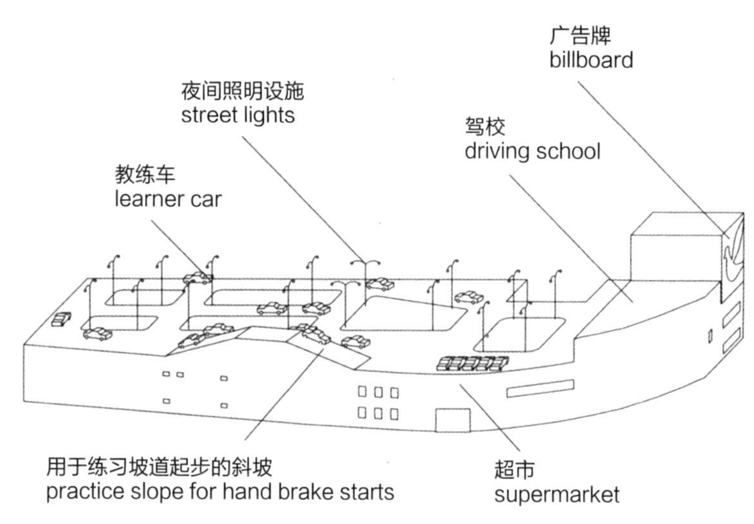

29

超级驾校

super car school

function: supermarket + driving school
site: Kanamachi, Katsusika-ku
- on top of the double layer supermarket lands a layer of driving school
- the site includes parcels of other people's property which could not be purchased
- the condition of the site, framed by the curve of the railroad,
is expressed directly in the extruded volume of the building
- the entry ramp is framed above by the practice slopes for hand brake starts

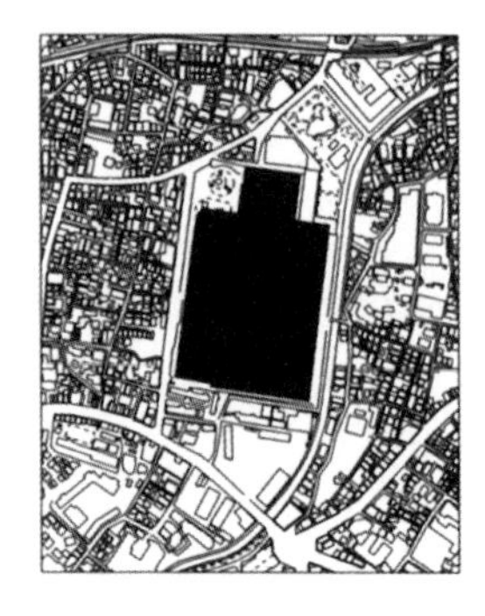

功能：污水处理设施+球场

位置：新宿区上落合

位于首都高速公路葛西立交枢纽附近○葛西污水处理厂的楼顶被改造成球场○共设有四个棒球场、一个足球场和一个橄榄球场○配有夜间照明设施○会有污水的异味○葛西地区还存在其他类似的建筑

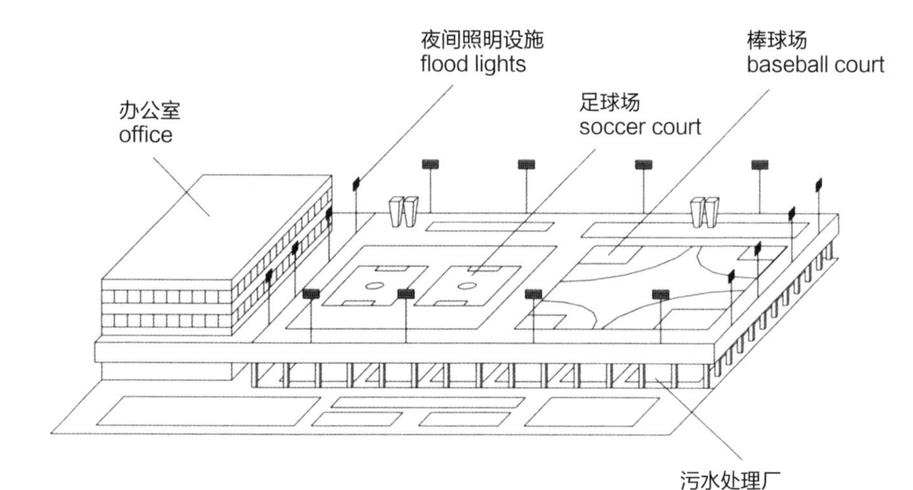

30

污水球场

sewage courts

function: sewage disposal plant + sporting facilities
site: Kamiochiai, Shinjuku-ku
- near the Kasai expressway junction
- the top of the Kasai sewage disposal plant is utilised for sports playing
- 4 baseball courts, 1 soccer court, 1 rugby court
- flood lights for night games
- scent of sewage
- another similar example in kasai

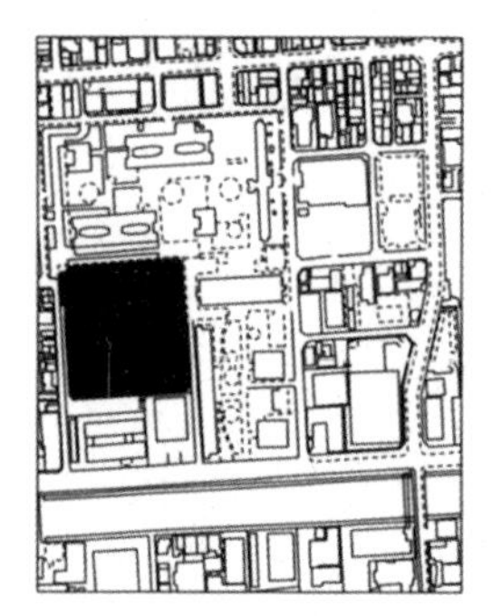

功能：蓄水池+球场

位置：江东区龟户

位于龟户天神社附近○横十间川沿岸的水务局，边长120米的正方形蓄水池上方建有棒球场和网球场○看起来就像邻接的公团住宅的中庭

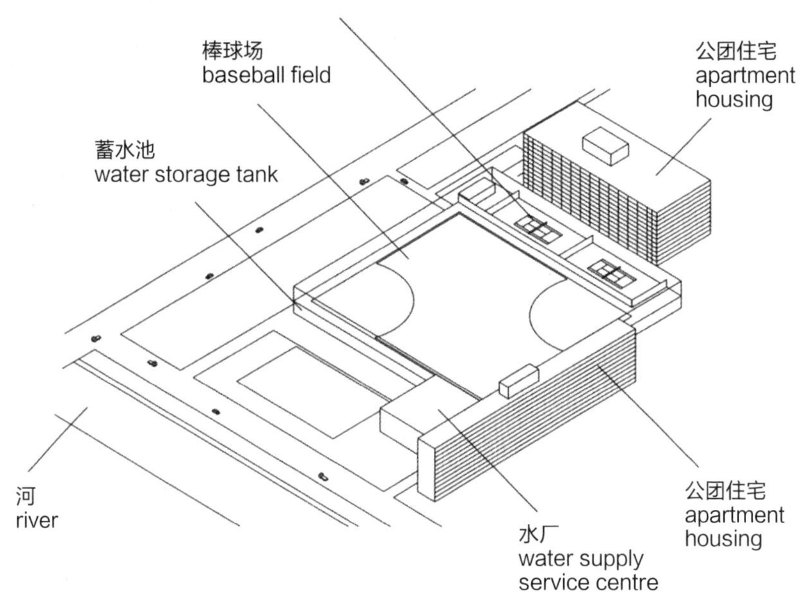

31

净水球场

supply water courts

function: water treatment plant + sporting facilities
site: Kameido, Koto-ku
- near Kameido Shrine
- water treatment plant along Yokojukken river
- 120 sqm surface area of water storage tank covered
by a baseball field and tennis courts
- appears to be a courtyard for the surrounding housing blocks

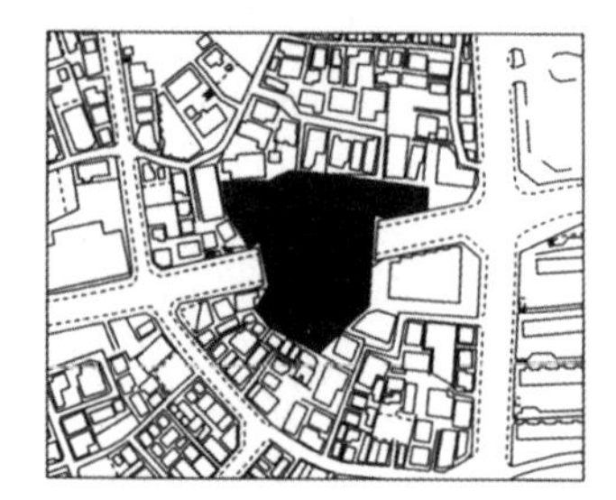

功能：墓地+道路

位置：涩谷区千驮谷

位于杀手街（外苑西路）和体育馆街交会的十字路口附近○道路从墓地下方经过○别名“幽灵隧道”○已故的学者池田贵族认为该地可以通灵

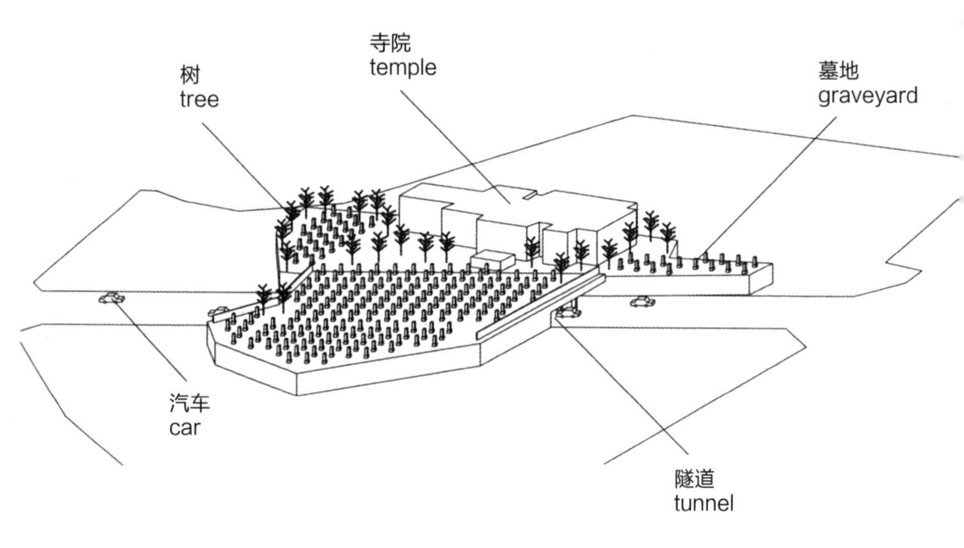

32
墓地通道
graveyard tunnel

function: graveyard + road
site: Sendagaya, Shibuya-ku
- around the area of the crossing of Killer street and Stadium street
- the road slices under the temple graveyard
- nicknamed 'ghost tunnel'
- the medium Kizoku Ikeda has named this a psychic spot

功能：寺院+店铺

位置：台东区上野

位于与山手线高架桥邻接的阿美横町的一角○寺院就建在阿美横町商店街的一块人工地基上○寺院的钟楼被各种招牌包围○店铺的招牌看起来就像是飘扬在寺院周围的幡○整条商店街上的店铺也兼作参拜道边的货摊

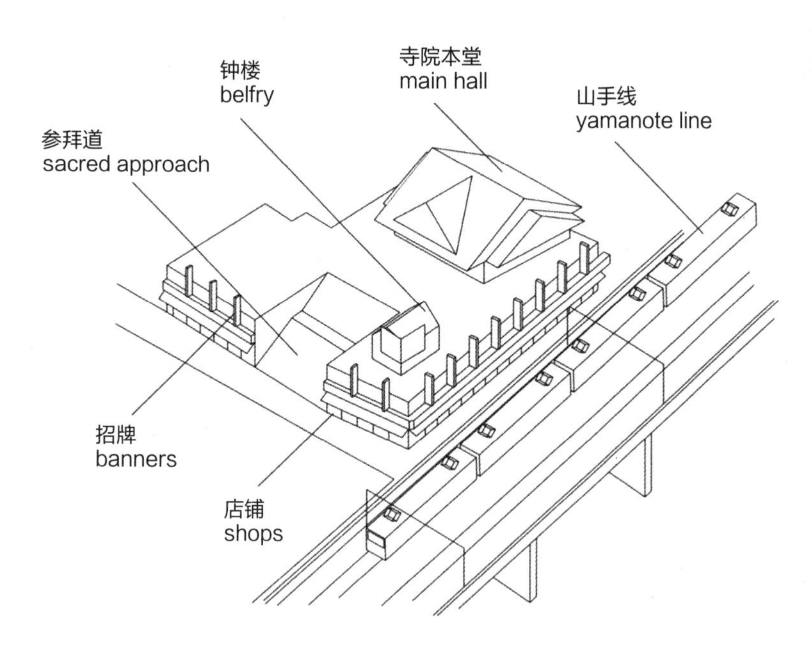

33

阿美横町空中寺院

ameyoko flying temple

function: temple + shops
site: Ueno, Taito-ku
- the block faces Ameyoko shopping street
and flies alongside the raised Yamanote train line
- the shrine precinct is established on artificial ground held up
by part of the Ameyoko shopping area
- the belfry is wrapped in signboards
- temple banners double as shop signage
- stalls traditionally aligning the sacred approach wind under the shrine precinct

二木の菓子
浜屋食品 株式会社
株式会社 大東商店
株式会社 大津屋商店
浜屋食品
浜屋食品㈱
大東商店
ダイマス
コクミン
ドラッグ
くすり
アメ横

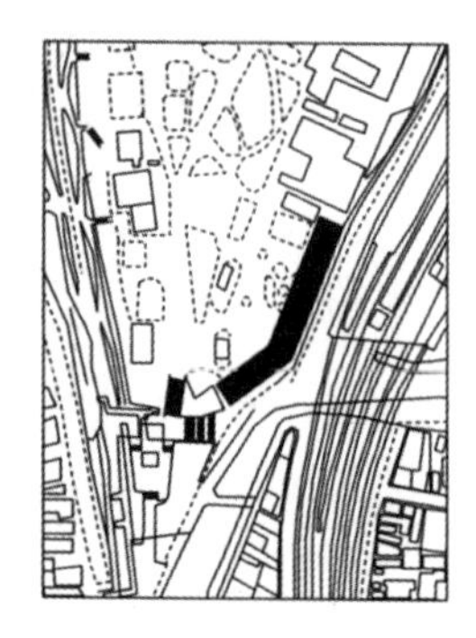

功能：百货商店+电影院+公园

位置：台东区上野公园

靠近山手线上野站○约130米长的挡土墙紧靠上野的山○百货商店的屋顶与上野公园相连，上面还有一尊西乡隆盛的雕像

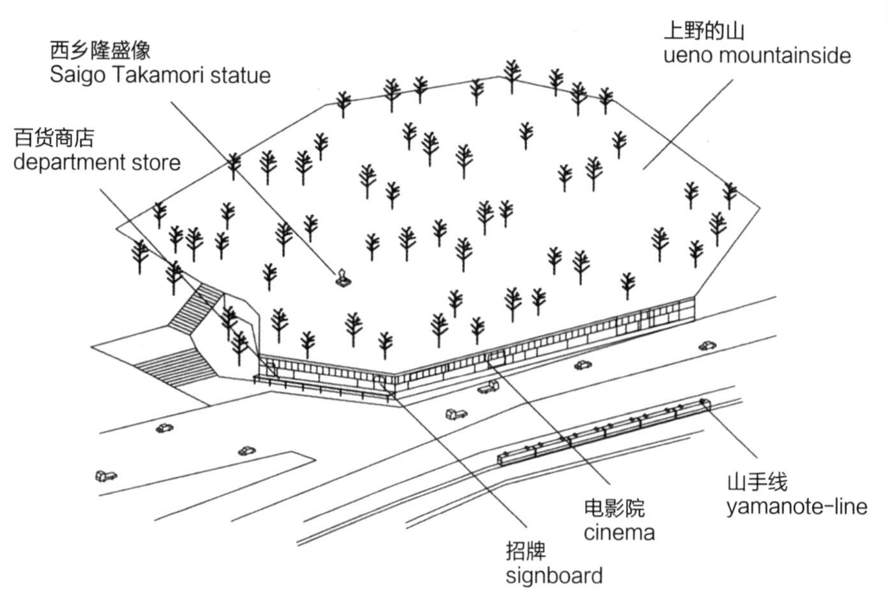

34

商店崖

shopping wall/mall

function: shops + cinema + park

site: Uenopark, Taito-ku

- near Ueno station on the Yamanote line
- 130m of architectural retaining wall holding back Ueno mountainside
- a statue of Takamori Saigo on the rooftop cum Ueno park ground

ライオン
P→上野パーキング
じゅらく
聚楽

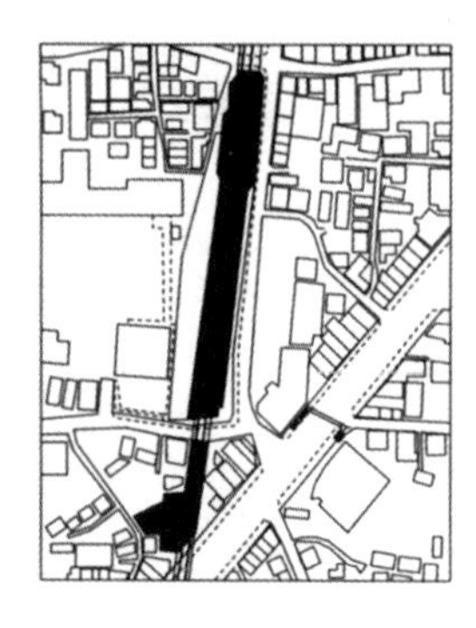

功能：博物馆+车站

场所：墨田区东向岛

东武铁道交通博物馆位于东武伊势崎线东向岛站下方○建筑整体呈长条状，对于展示铁道车辆再合适不过了○参观者还可以透过月台下的窗户看到现在正在使用的铁道

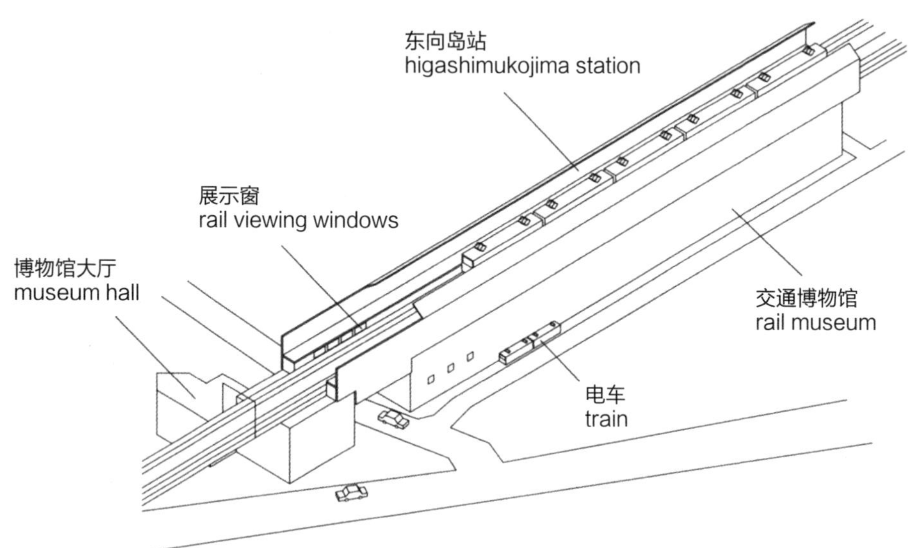

35

铁道博物馆

rail museum

function: rail museum + rail station

site: Higashimukojima, Sumida-ku

- museum of Tobu railway lines under Higashimukojima station of the Tobu Isezaki line
- the linear form of the site is ideally suited to display the length of each carriage
- windows underneath the platform-in-use show a close up view of the rai ls themselves

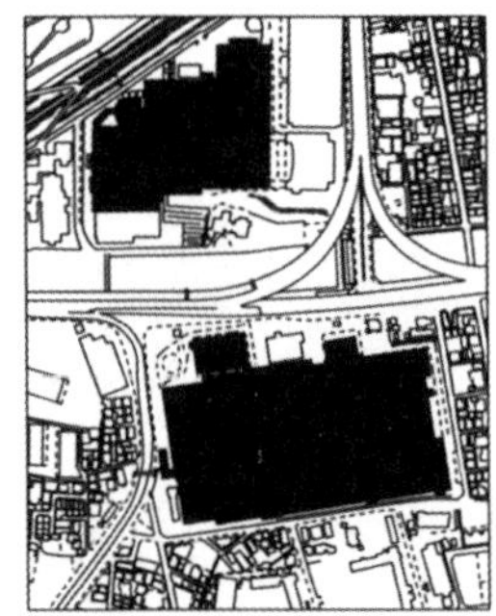

功能：污水处理设施+公园

位置：葛饰区小菅

位于首都高速公路小菅出口附近○首都高速公路和河川把污水处理厂隔离在两岸○其中一侧的污水处理厂为配合首都高速公路的弯道转向，整体上设计成了雁阵的形态○另一座污水处理厂的楼顶上设有五片网球场和一处带池塘的日式庭园○现场闻不到污水的臭味

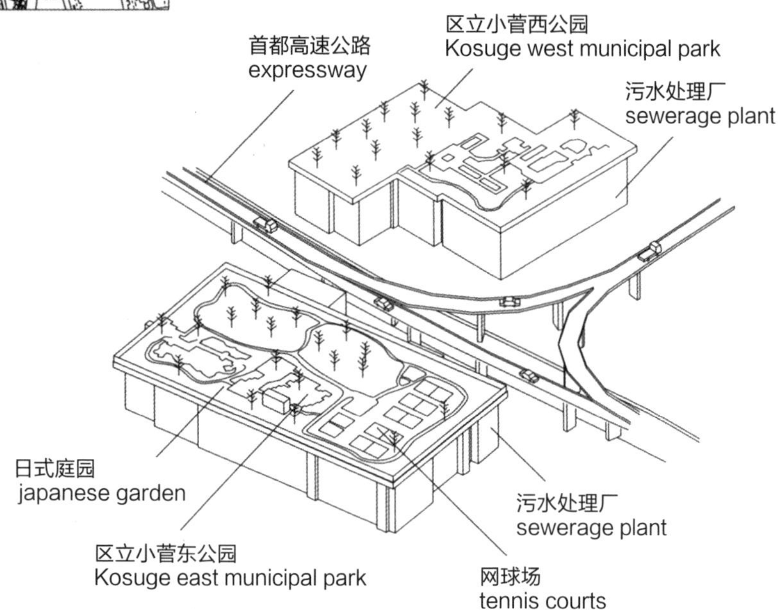

36

双子污水花园

twin deluxe sewerage gardens

function: park gardens + sewerage plant
site: Kosuge, Katsushika-ku
- near the metropolitan expressways' Kosuge junction
- the expressway route and river divide the plant into 2 parts
- one side adjusts to the curve of the expressway corner by its zig zag form
- the roof terrace houses 5 tennis courts
and a traditional Japanese garden, including a small lake
- no sense of sewage

功能：泳池+餐厅+游戏厅+大厅+停车场
位置：足立区西新井
又被称作“东京之海”○住宅区里的一个巨大泳池○一层是游戏厅、停车场、餐厅，楼顶是泳池和栽种了许多植物的庭园○为了追求更加刺激的滑行体验，滑道每年都在扩建

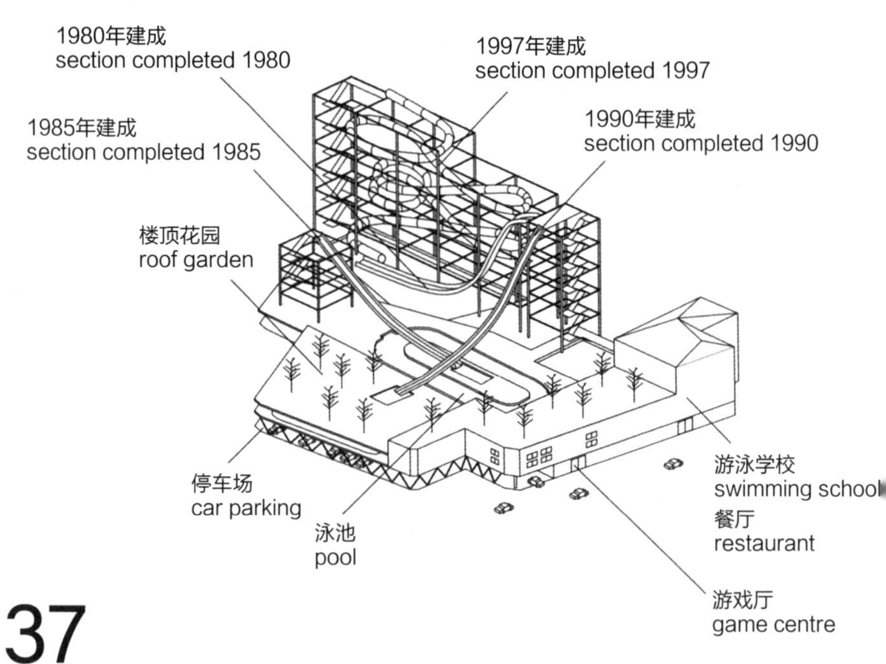

37

增殖滑道楼

proliferating water slides

function: play pools + restaurant + game centre + hall + parking
site: Nishiarai, Adachi-ku
- "Tokyo Marine"
- the enormous pool towers over the residential area
- the ground floor contains game centre, car parking and restaurant,
the roof terrace contains pool facilities and planted garden
- the growing number of slides are becoming more and more sensational each year

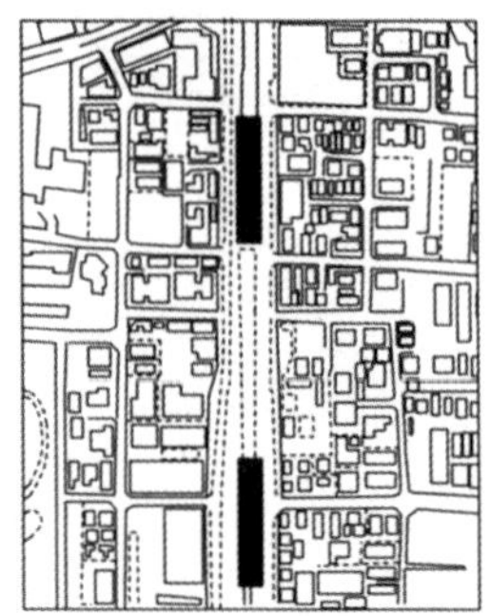

功能：隧道通风塔

位置：杉井区井草

位于环状8号线的井荻隧道附近○位于环状8号线上的两座巨大通风塔隔着西武新宿线遥遥相望○通风塔内还配有机械室、管理室、业务办公室等空间○足以与巴黎旺多姆广场方尖碑相提并论（？）的新视线落点

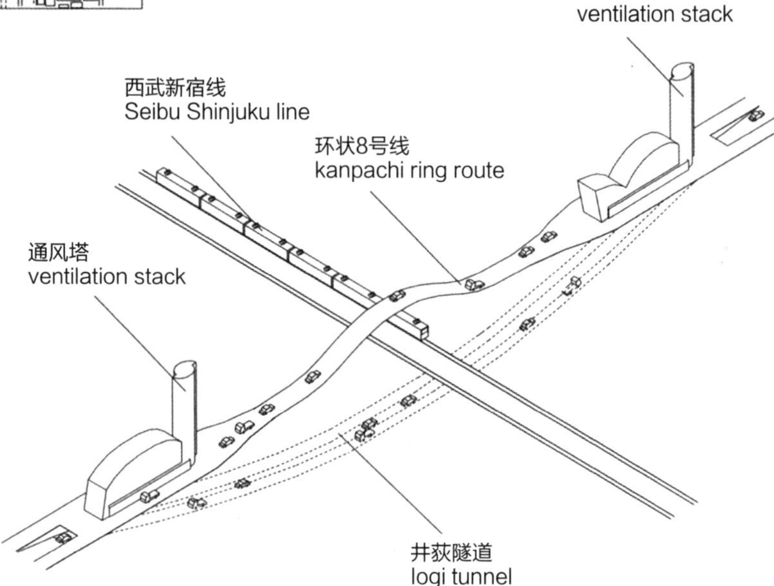

38

通风方尖碑

ventilator obelisk

function: tunnel ventilation
site: Igusa, Suginami-ku
- attached to Kanpachi-Iogi tunnel
- the obelisk-stacks face each other across the Seibu Shinjuku line, and stand resolutely in the middle of traffic
- machine rooms and offices are linked to each obelisk
- a new kind of landmark rivalling the obelisk of Vendome Place in Paris

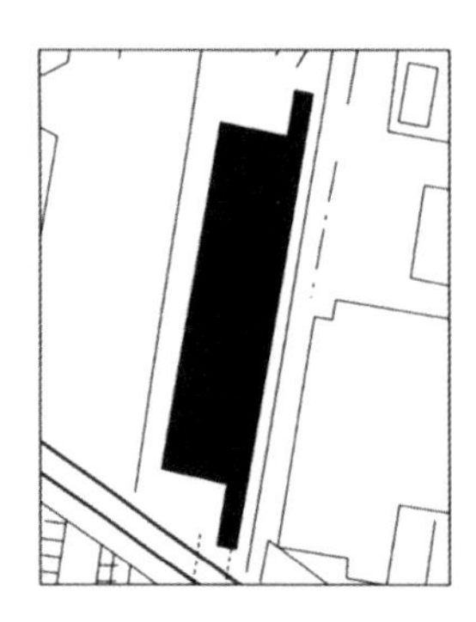

功能：车站+集合住宅+车库

位置：松户市幸谷

位于私营铁路流山线幸谷站和坂川之间○L字形集合住宅一层的走廊同时也是车站的月台○与走廊邻接的一层则充当出租车的车库○出了公寓大门就是坂川防波堤步道

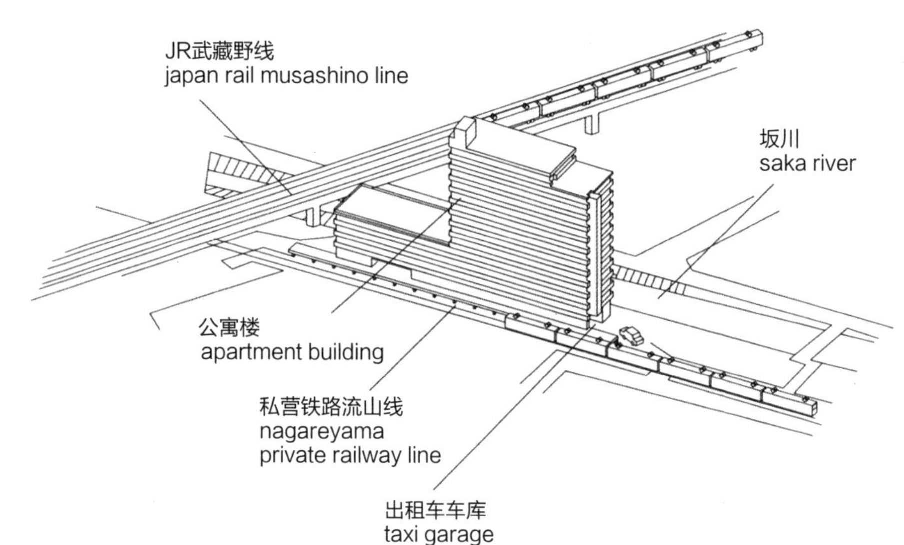

39

车站上的家

apartment station

function: train station + apartments + taxi garage
site: Kaya, Matsudo-shi
- sandwiched in between Kaya station of the Nagareyama line and Saka River
- the housing forms an L-shaped section,
the train station platform appears as the ground floor balcony
- the ground floor is a garage for taxis
- the concrete promenade to the river is continuous with the building's entry

幸谷駅
KOYA STATION

功能：铁道+集合住宅

位置：八王子市东浅川町

位于京王线高尾站附近的高架桥下○私营铁路公司的员工宿舍○共43户，建筑总长约300米○规格统一的住宅被依次插入高架桥的桥墩之间

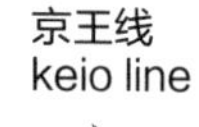

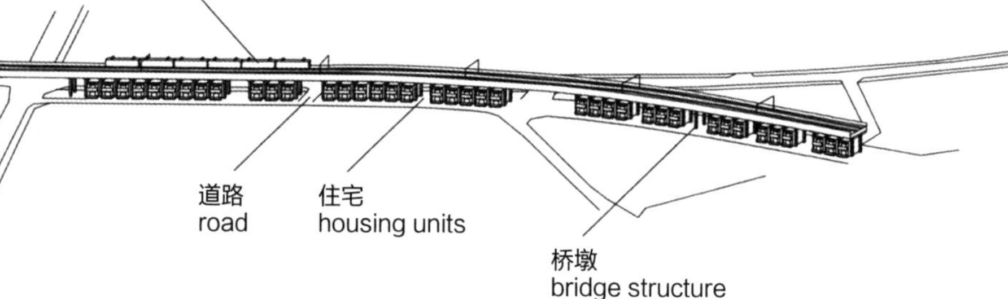

40

蜈蚣住宅

centipede housing

function: railway + housing units

site: Higashiasakawa-cho, Hachioji-shi

- underneath raised railway structure near Takao station, Keio line
- utilised as Keio company housing
- 43 housing units over approximately 300 metres length
- the housing unit volumes are inserted to match the structural span

功能：车检+维修销售+保险+涂装+钣金+公营停车场+住宅

位置：板桥区大山金井町

位于首都高速公路下方的中仙道沿线○转车台被各式与汽车相关的建筑物包围○停车塔的立面是汽车经销商用于发布消息的电子显示屏○所有建筑物一同组成了一个汽车社区

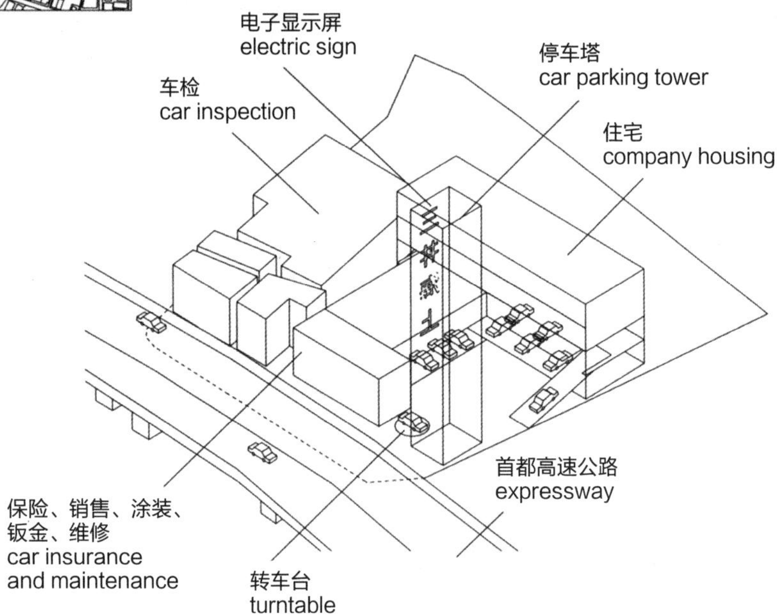

41
汽车村
vehicular village

function: car inspection + maintenance + show room +
insurance+ public parking + company housing
site: Ooyamakanaimachi, Itabashi-ku
- alongside Nakasendo road and the expressway overhead
- the various buildings of the village gather around the turntable courtyard
- the tower runs an electric message from the showroom down the facade
- auto community

定期点検
千代田自動車
でも 調子が悪い
POCARI SWEAT
CAR

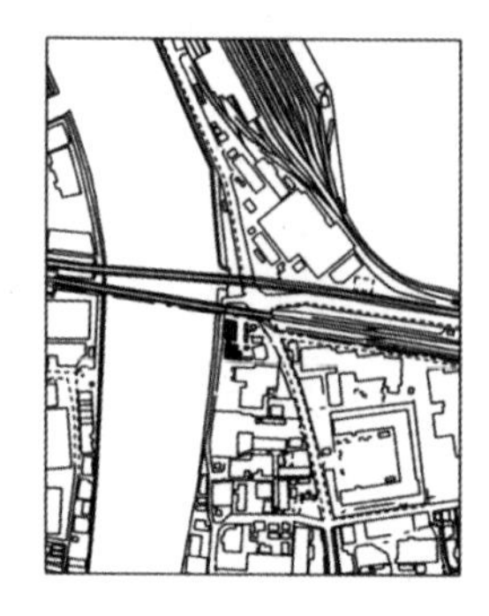

功能：潜水学校

位置：荒川区南千住

位于隅田川沿岸，可以从常磐线的电车上望见○蓝色的圆筒状水塔上设有螺旋状楼梯，与箱型的教室和仓库相连○储水罐外侧开有几扇圆形亮窗，从那里可以眺望街道○顶部的潜水头盔是整座建筑物的亮点○潜水学校由一家潜水器材公司经营

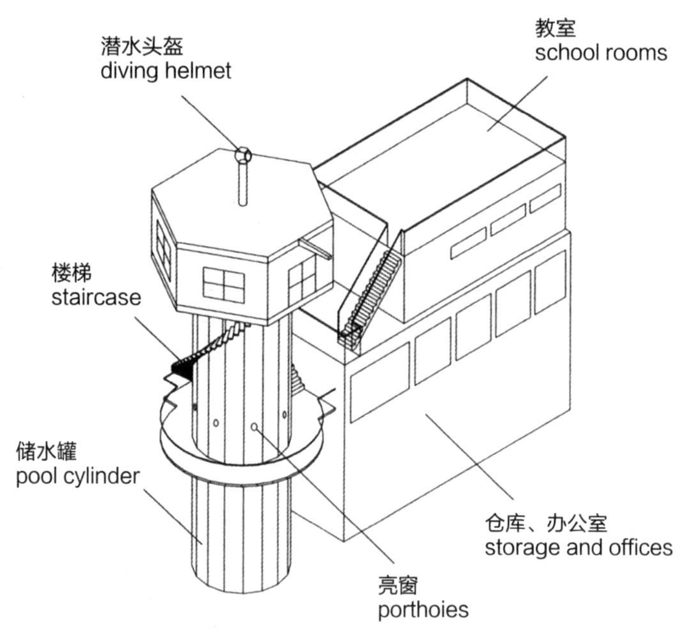

42

潜水塔

diving tower

function: diving school

site : Minamisenju, Arakawa-ku

- alongside Sumida River, visible from the Joban railway line
- a staircase spirals around the blue cylinder tower pool
and docks into the school rooms and storage block
- portholes for looking out towards the city
- a diving helmet is raised as a pinnacle for the tower
- management by diving tank producing company

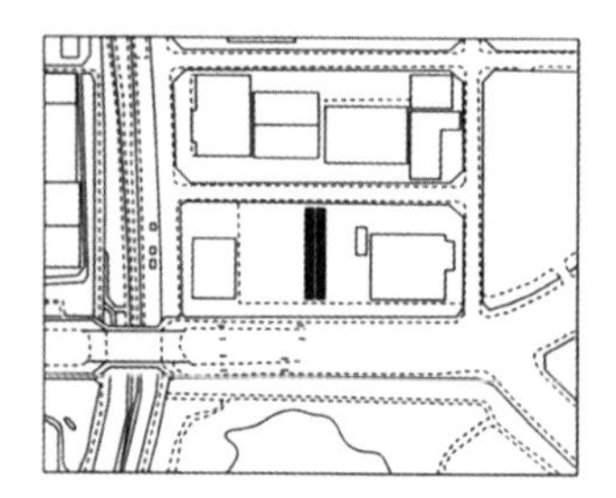

功能：立体集装箱放置区

位置：大田区东海

位于都立东京港野鸟公园前○用于放置集装箱的货车底盘共十层，堆叠成立体的形态○列与列之间用颜色区分○垂直放置集装箱省出的空间被用作停车场

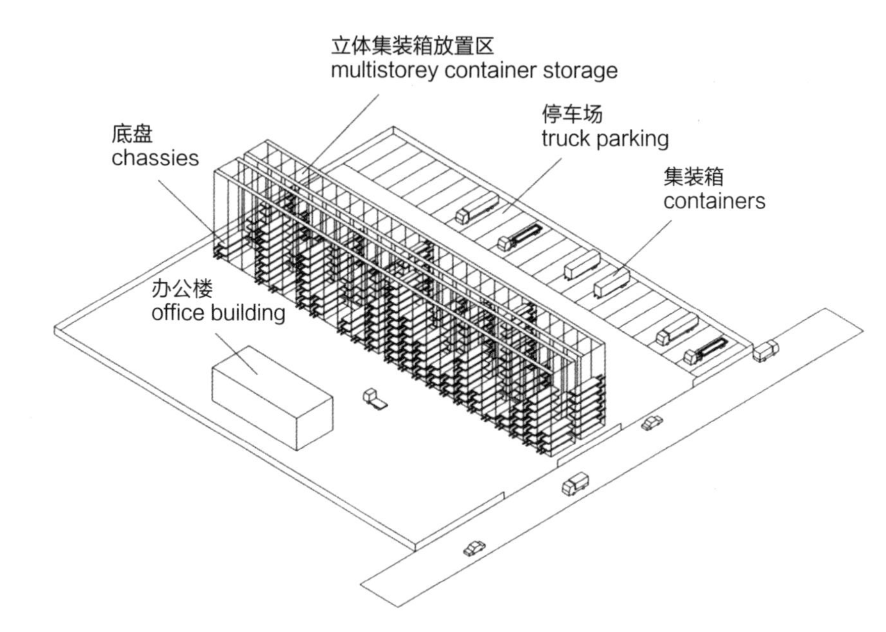

43

底盘公寓

chassis apartments

function: multistorey chassis storage + truck parking
site: Takai, Ota-ku
- in front of the municipal Tokyo Bay wild bird sanctuary
- truck chassis are layered 10 levels high to be packed into a solid block
- chassis are colour coded in rows
- the ground space gained by raising
the containers is utilised for truck parking

SUZUE
FUSO

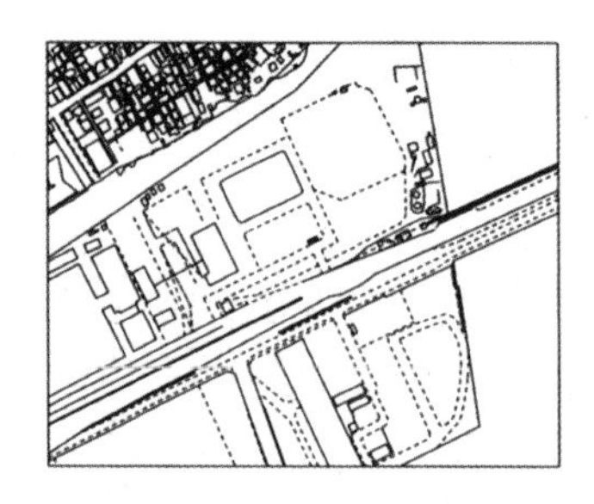

功能：办公室

位置：川崎市鹤见区生麦

位于产业道路鹤见大桥附近○事务所由四个堆叠在集装箱放置区入口的集装箱改造而成○蓝色集装箱上用斜体书写的三色文字是整座建筑的标志○集装箱、铝窗、钢架楼梯、天线、太阳能板、电线等元素在这里意外相聚

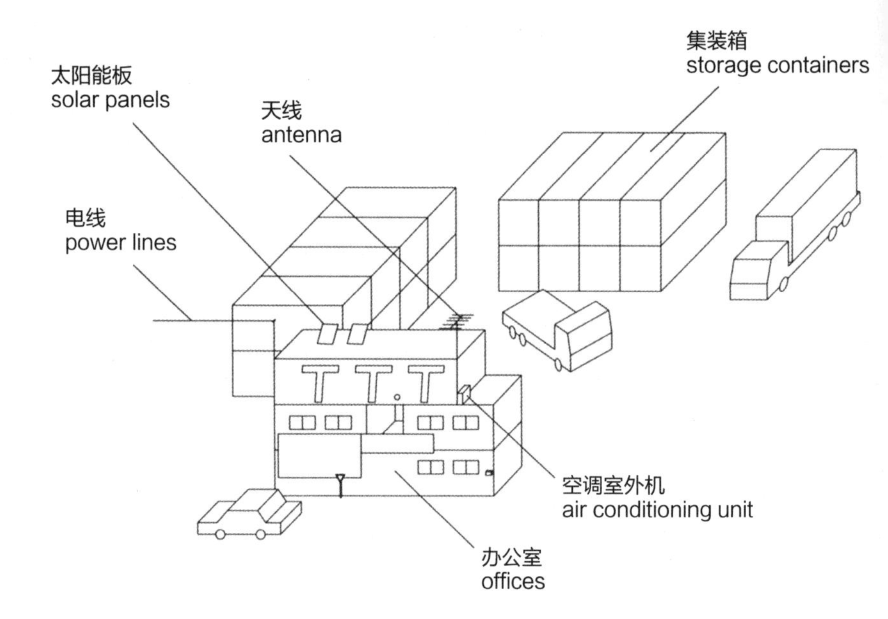

44

TTT（乐高办公楼）

TTT (lego office)

function : container offices + container stockyard
site: Namamugi, Tsurumi-ku, Kawasaki
- near Tsurumi Ohashi bridge, Sangyo road
- office building made up of 4 containers stacked at the entry of the container yard
- coloured italic text on a blue container body makes
a tricolour identity for the company
- unexpected coming together of storage containers,
aluminum sashes, steel stairs, antennas, solar panels, power lines

T.T.T
TANEY (TOKYO)
TANK TERMINAL
TEL. 045 510 2500
FAX. 045 510 2475
安全第一

功能：上越新干线+神社

位置：北区赤羽台

位于埼京线、东北新干线和上越新干线赤羽站附近○神社的下方是隧道○神社的后方是私立小学和中学○参拜时经过的坡道和新干线的铁路平行○乘坐新干线上下班，不知不觉间就完成了百次参拜

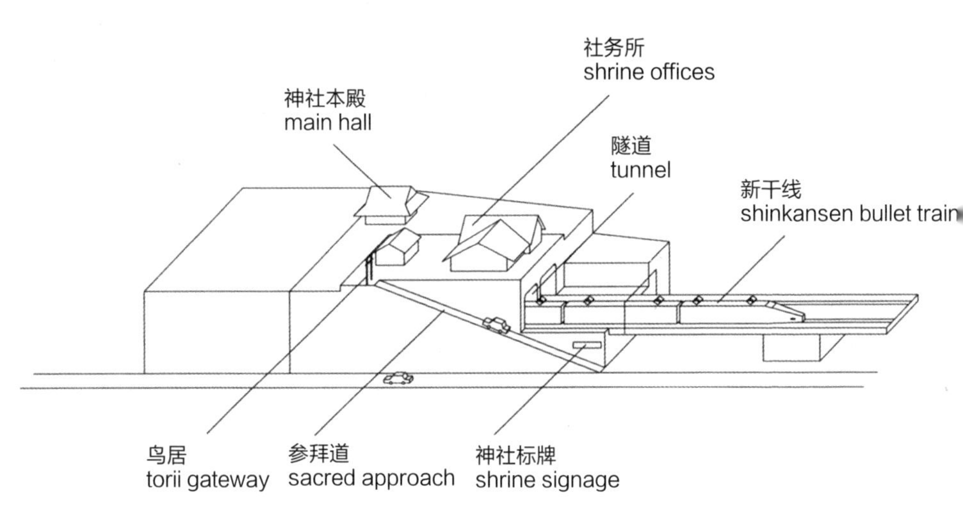

45

隧道神社

tunnel shrine

function: shinkansen + shrine

site: Akabanedai, Kita-ku

- near Akabane station of Saikyo and Shinkansen lines
- tunnel under the shrine precinct
- further into the hillside is a private primary and middle school
- the sacred approach slopes parallel to the Shinkansen route
- Shinkansen commuters have the opportunity to make daily prayers without realising it

功能：寺院+集合住宅

位置：横滨市保土谷区台町

位于东横线反町站附近○毗邻金比罗神社○去寺院参拜必须经过公寓内部的楼梯和公寓楼顶的桥○公寓楼顶相当于寺院前庭，铺设了防水布以代替粗砾石○对称放置的鸽子小屋（通风设备）犹如寺院的灯笼

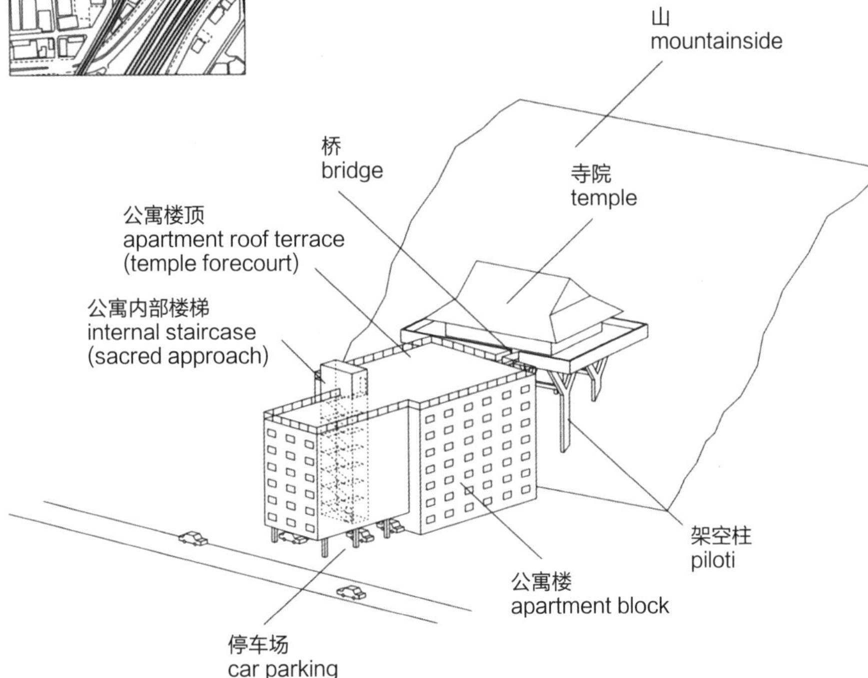

46

公寓山寺

apartment mountain temple

function: buddhist temple + private apartment housing
site: Daimachi, Hodogaya-ku, Yokohama-shi
- near Tanmachi station of Toyoko train line
- Konpira shrine is adjacent
- the religious approach passes the internal staircase of the housing, across the apartment roof and bridge to the main temple hall
- the forecourt of the temple is not pebbles or gravel, but the waterproof sheeting of the apartment roof terrace
- ventilators aligned in parallel on the roof terrace are like lanterns

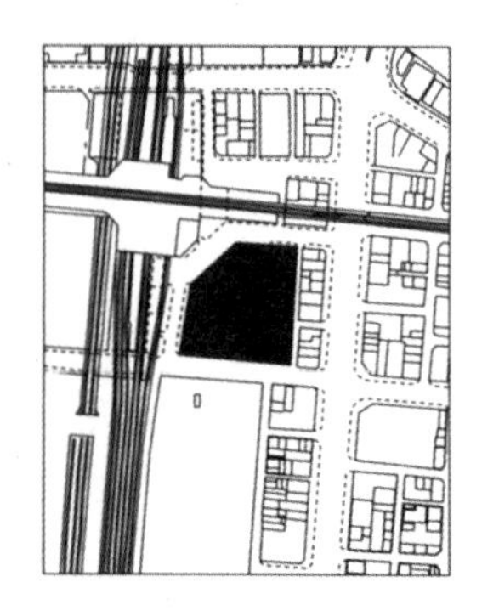

功能：车站前广场+献血站
位置：千代田区外神田
位于秋叶原站前○广场被秋叶原布满霓虹灯招牌的大楼围绕，为三对三篮球、滑板等城市运动设立了专门区域○献血站安静地待在广场的一角，广场一侧和公路一侧都设有入口○渴求着运动的年轻人那新鲜的血液

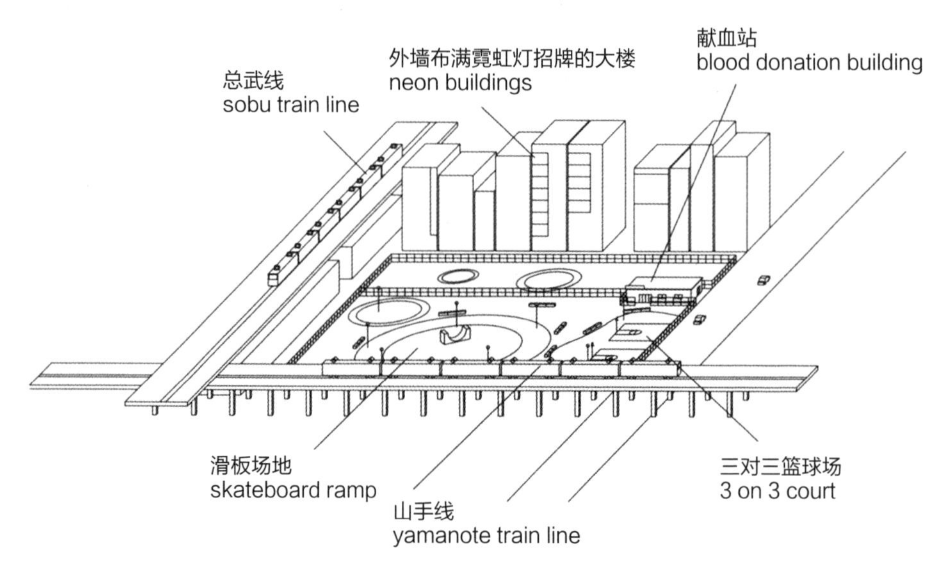

47

吸血公园
vampire park

function: public square + blood donation
site: Sotokanda, Chiyoda-ku
- in front of Akihabara station
- neon buildings of Akihabara surround open space containing urban sports such as skateboard ramps and 3 on 3 courts
- the blood donation building sits innocently in the square with entries to both the square and the street
- desire for young and fresh blood

冷暖
専門店
AASPRO
アンテナ
BS·CS
TV·FM
トミヒサムセン
VAIO
Nishimura
おんころ·パソコン·タワー
PC98-NX
誕生
Let'snote
TOSHIBA
ダブルウインドウ
坂口電熱はここです
光LAN
ネットワーク
AISAN
愛三電機
創業50年
aisan.co.jp
光ファイバー
光通信機材
ギガスイッチ
100M
インターネット
セキュリティ
UPS
無線LAN
献血ルーム

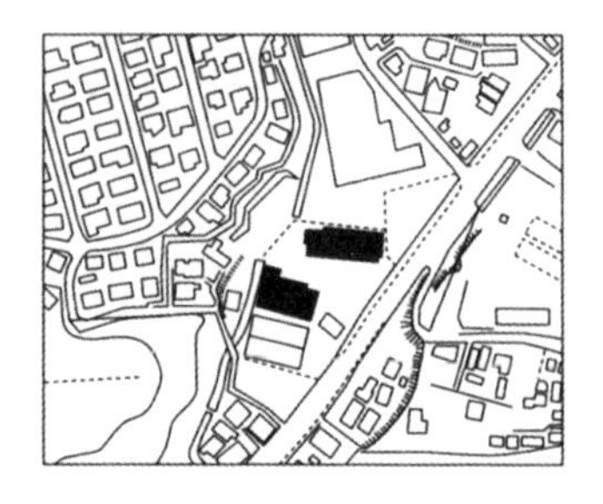

功能：起重机公司房屋
位置：横滨市南区平田町
位于国道1号线沿线○塔式起重机公司的仓库兼办公室○五颜六色的桁架堆叠在铁架上○因为公司经营不善，这些塔式起重机出勤机会寥寥

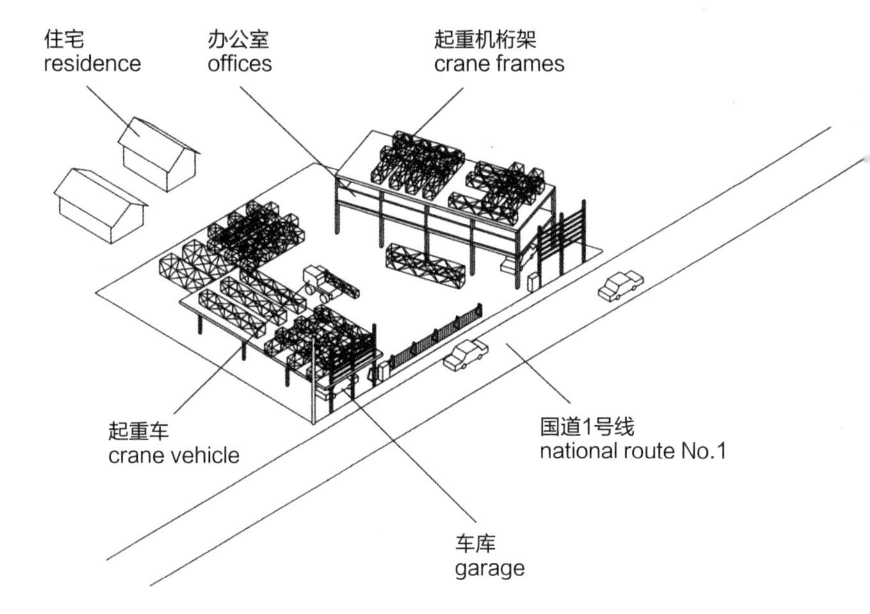

48

起重机架

crane shelves

function: crane company office building
site: Hiratamachi, Minami-ku, Yokohama-shi
- alongside National Route no.1
- crane storage and office facilities for crane company
- colourful truss frames are stacked on steel shelves

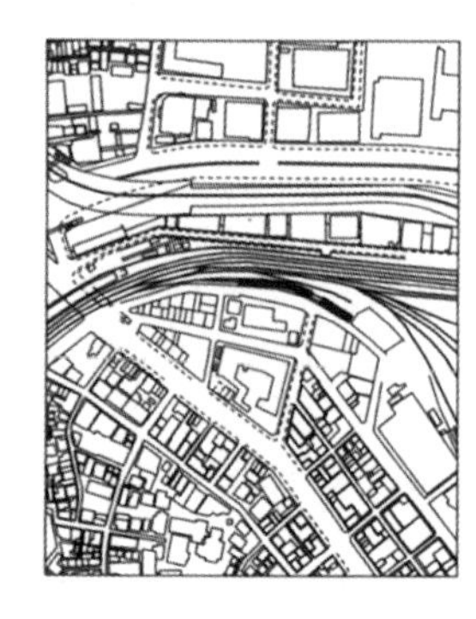

功能：旧货运线高架桥+泥瓦工厂

位置：千代田区饭田桥

位于通往旧货运线饭田町站高架桥下的一家泥瓦工程公司○货运线停运后，公司在高架桥上加盖了楼房○与前方柏油公路相连的高架桥桥洞用于储藏土、沙子、砂砾、石头等骨料○建筑物整体呈弓形，公路从中间横穿而过

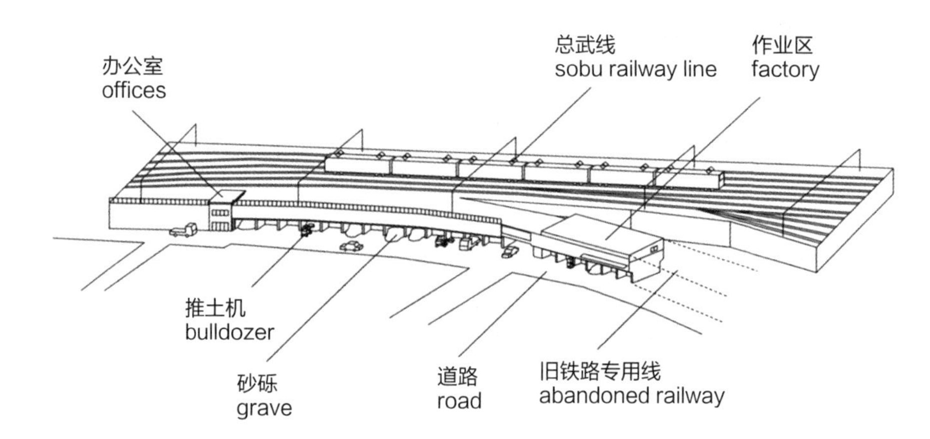

49

幽灵铁道工厂

ghost rail factory

function: abandoned railway + plaster factory
site: Iidamachi, Chiyoda-ku
- plaster factory under the structure of the disused branch line for the old freight train to Iidamachi station
- following the discontinuation of the freight line, the company has extended by building the factory on top of the rail track space
- the structural span makes pockets for storing earth, sand, gravel, stones etc, facing directly onto the asphalt road
- a road cuts through the bow shaped building

功能：公寓+挡土墙

位置：川崎市麻生区细江

位于小田急线读卖乐园站附近○为支撑被削平的山而建起的挡土墙与公寓之间通过井字梁连接，所以两者看起来像是一个整体○如此形成了挡土墙支撑山体，公寓又支撑挡土墙的力学连锁结构○公寓外部的走廊被刷上了油漆，看上去有别于一般的土木结构物○无数的井字梁纵横交错在公寓和挡土墙之间的外部空间内

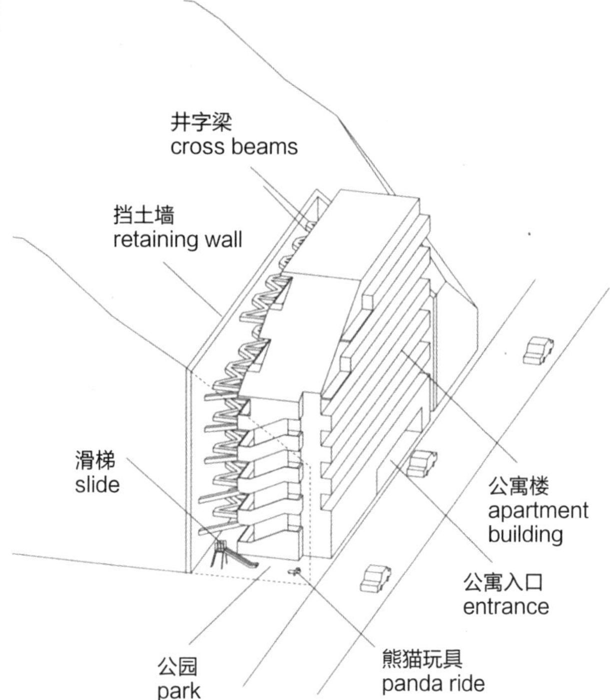

50

挡土墙公寓

retaining wall apartments

function: apartment building + retaining wall
site: Hosoe, Asou-ku, Kawasaki-shi
- near Yomiuri Land station of Odakyu railway line
- the retaining wall and the apartment building
become one through the criss-crossing structure
- the wall holding back the mountain joins forces with the weight of the building
- an attempt to separate architecture and engineering
by painting the external balcony but not the criss-cross structure
- a strange space filled with crossing structure appears

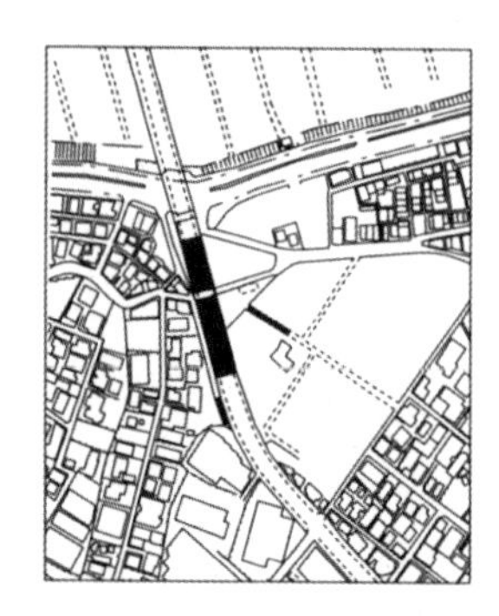

功能：公园+高架桥+儿童收容所

位置：港区南青山

位于青山桥下，青山桥连接起被青山墓地包围的麻布谷和六本木的台地○桥柱的配色是勒·柯布西耶式的，现代主义风格的儿童收容所建于桥柱之间○与其邻接的区立儿童乐园和青山墓地就如同儿童收容所的配套设施○独占一片城市中心少有的静谧和葱茏

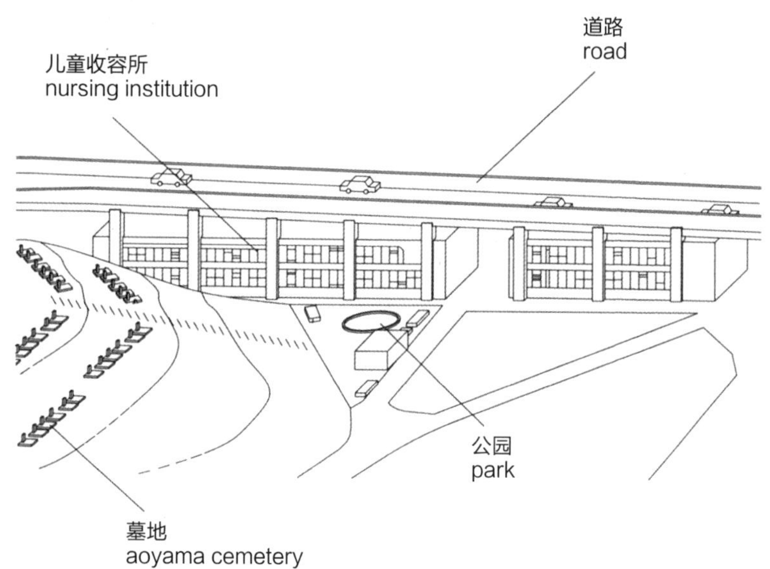

51

桥洞里的家

bridge home

function: children's home + park+ raised roadway
site: Minami-aoyama, Minato-ku
- under the Aoyama bridge which spans over the valley of Aoyama cemetery
- modern architecture of Corbusian colours makes use of bridge structure as piloti
- Aoyama cemetery and the park neighbouring the home are like grounds to an estate
- the home monopolises this serenity and abundant greenery
which seems unthinkable in the middle of the metropolis

功能：农户+农田+蔬菜摊
位置：练马区早宫
受城市扩张影响，周围的农田都被改造成了住宅用地，农场所在之处仿佛被拔掉了一颗牙齿○面向道路一侧的蔬菜摊以低廉的价格出售农田里收获的新鲜蔬菜○蔬菜摊背后是农家的主屋，两者由田间小道连接○零运费的超微型循环经济的范例○田地边有祖先的墓

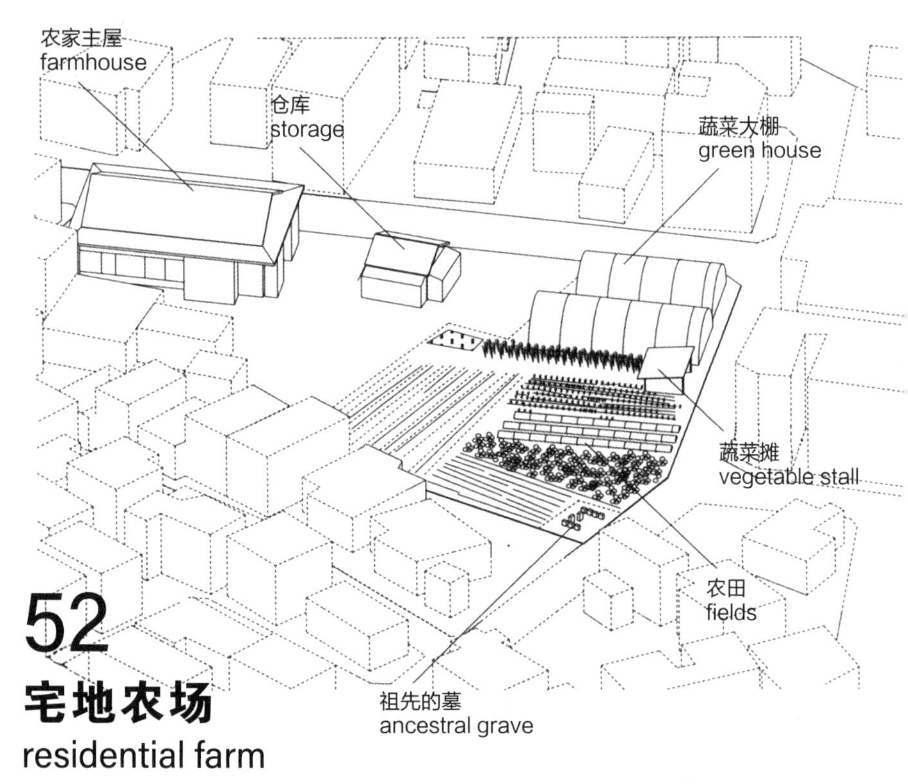

52
宅地农场
residential farm

function: farmhouse + fields + vegetable stall
site: Hayamiya, Nerima-ku
- a suburbia developed in random pockets of farming land
- on the road side of the fields is a stall selling
fresh vegetables direct to the customer at low prices
- the farmhouse and stall are connected by a thin foot path through the fields
- a super-shortcut economic cycle needs zero transportation time or cost
- the tradition of the ancestor's grave in the corner of the fields
is retained despite surrounding residential development
- the ultimate situation for the direct sale of produce

功能：快递物流中心

位置：埼玉县新座市

毗邻关越公路练马收费站○把关越公路上大型货车的货物分流到小型货车上的物流中心○一层是用于分流作业的快递集散区，二层是物流公司办公室，楼顶是货车的停车场○快递随着大分贝流行歌曲的节奏，被送上传送带

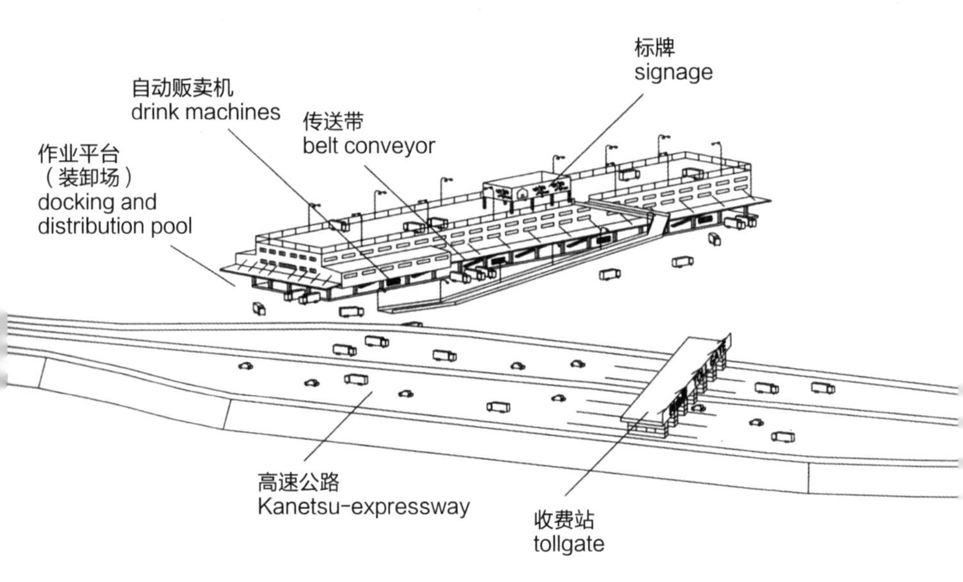

53

物流立交枢纽

dispersal terminal

function: courier station
site: Niizashi, Saitama-ken
- next to the Nerima tollgate of the Kanetsu expressway
- a station for transferring large packages
on large trucks into small packages on small trucks
- ground floor docking and transfer pool,
middle floors administration, rooftop truck parking
- packages conveyed around the transfer pool,
riding on the rhythm of radio hit stations

功能：停车场+公寓+家庭餐厅+高尔夫球练习场
位置：板桥区樱川
位于川越街道沿线，建在高尔夫球练习场的用地上○一层是停车场，二层是家庭餐厅和高尔夫球练习场，三层是高尔夫球练习场和住宅○买下这里的公寓就附赠餐厅和高尔夫球练习场○高尔夫球练习场四周加装了挡网，不过球还是会直接打到公寓走廊○吃饭+打球，正是爸爸们理想的周末生活

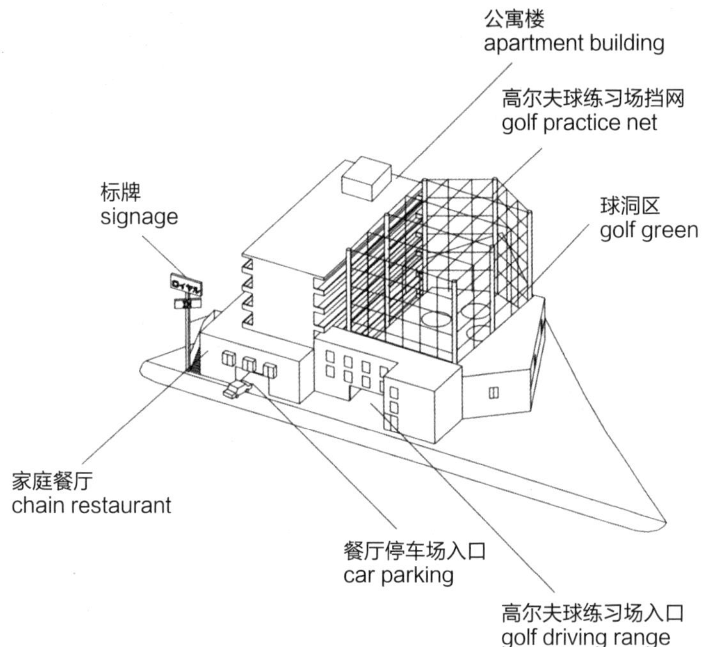

54

皇家高尔夫公寓

royal golf apartments

function: 'royal host' chain restaurant +
golf practice range + apartment building + car parking
site: Sakuragawa, Itabashi-ku
- next to Kawagoe road, previously a ground level golf practice range only
- ground floor car parking, middle floors 'royal host'
and golf practice range, upper floors apartments
- housing with the benefit of a spacious netted forecourt and optional restaurant
- a fathers' weekend package is available inside metropolitan Tokyo
- beautiful view through the nets of the golf practice range from the external corridors

功能：立体停车场+餐厅

位置：中央区银座

位于松屋大道沿线的“东京停车大厦”○一、二层是餐厅，一层还有入口通往带转车台的停车场○因为地价高昂，所以取消了停车坡道○位于建筑中央的电梯旋转上升，呈放射状把车运往各个楼层○白色混凝土和铝制框架让整座建筑看起来像是普通的办公楼

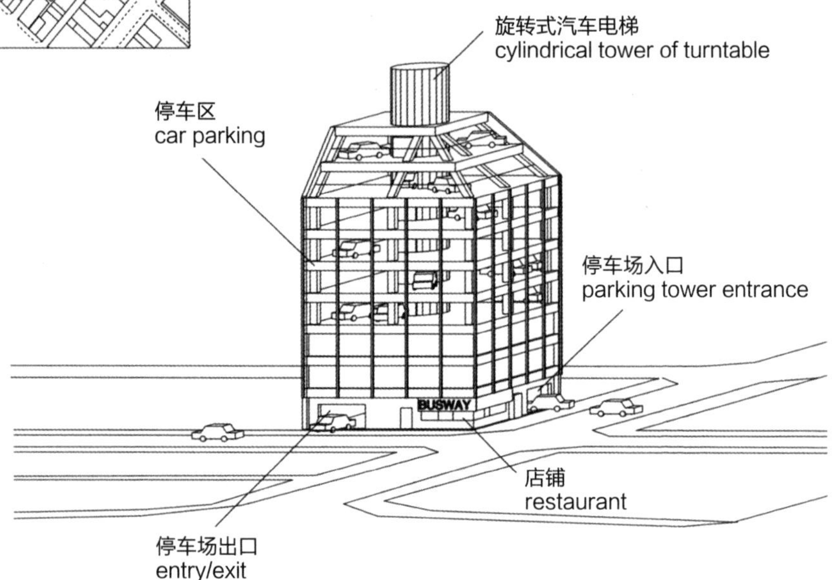

55

汽车办公楼

car parking office

function: parking tower + restaurant
site: Ginza, Chuo-ku
- the original 'Tokyo parking building' faces Matsuya street
- the lower 2 floors are utilised by restaurants except
for the carpark entry leading to a turntable tower
- due to the extremely high price of land in Ginza,
there is no chance to waste space on circulation ramps
- the pure white concrete and aluminum frames show off the cars inside
- the revolving lift is a highlight
- the final possible extent of the mechanisation of architecture, before the machine
completely takes over, as in more contemporary parking towers

SUBWAY

功能：立体停车场+庭园
位置：品川区西品川
园林公司的庭园和立体停车场重叠在一起○停车场的格栅平台配合庭园的轮廓修建○庭园里的树穿过停车场的格栅平台，高度超过了房顶○汽车、苗木和园林设备在一层共存○入口斜坡处安装了自动贩卖机，大家可以在此停车休息○隔壁就是园林公司的办公室和公司老板的私宅

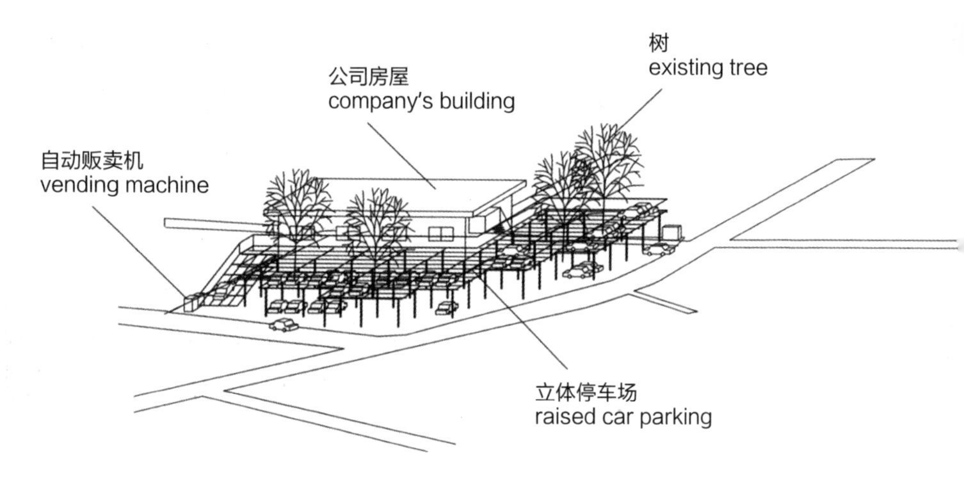

56

绿色停车场

green parking

function: multi-layer ramped car parking + garden
site: Nishi Shinagawa, Shinagawa-ku
- headquarters of a landscape company,
and company president's home beside a carparking building
- because the site used to be a garden, there still remains a stone wall and some trees
- keyaki and persimmon trees push up through the grating of the floor above
- a sparse building including much greenery
- drink machines cluster around the ramp entry to encourage taking a break

功能：汽车修理厂+汽车用品商店+停车场+保龄球馆

位置：练马区北町

位于环状8号线和川越街道的十字路口○车主可以将车直接开进二层的停车场○一层是汽车修理厂和汽车用品商店，二层是停车场，三、四层又是汽车用品商店，五层是游戏厅和保龄球馆○在修车期间，顾客可以购物、打保龄球消磨时间○清水混凝土上的黄色招牌和上方巨大的保龄球瓶是此处的标志

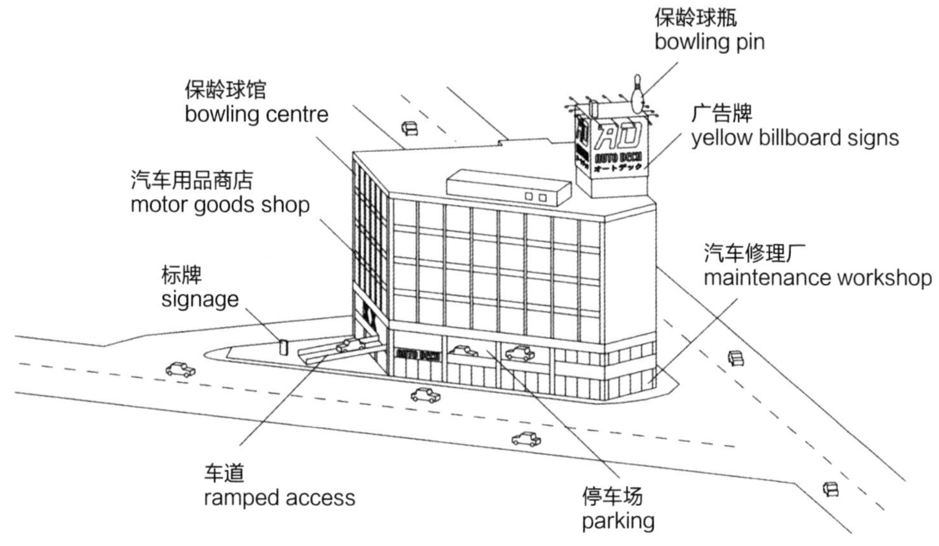

57

汽车百货商店

auto department store

function: maintenance workshop + motor goods shop +
store user parking + ten pin bowling
site: Kitamachi, Nerima-ku
- at the corner of Kanpachi ring route and Kawagoe road,
the building volume fills the extent of the site
- direct, drive-in, ramped access to second storey user parking
- car related activities are stacked on top of each other
- yellow billboards over undressed concrete facades,
a huge bowling pin is the crowning sign
- bowling in the spare time created by waiting for car repair work

AUTO TECH
オートテック
創業20周年記念セール
ビックテック
Enjoy Car Life!!
カーナビゲーション
オイル/バッテリー
ワックス/洗車用品
カーアクセサリー
タイヤ/アルミホイール
日本最大級の
カー用品専門店
日本最大級のカー用品専門店
ビックテック
AUTO TECH
AUTO TECH

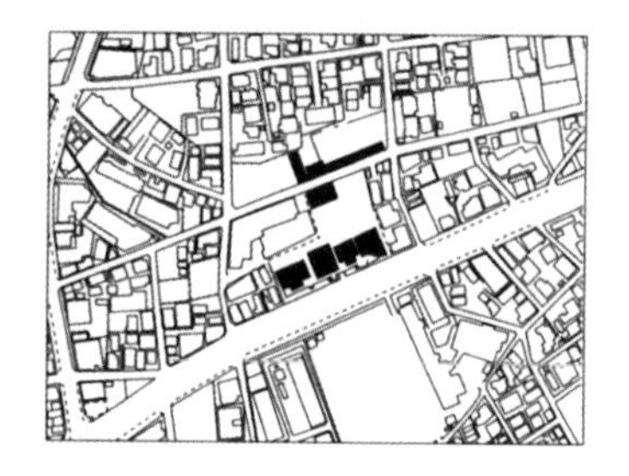

功能：家庭餐厅+体育用品商店+停车场
位置：丰岛区南长崎
位于目白大道沿线○三间家庭餐厅和一家体育用品商店依次排列在一片人工地基上○与隔壁的高尔夫球练习场共用一层的停车场○西餐、中餐、日料任你选择

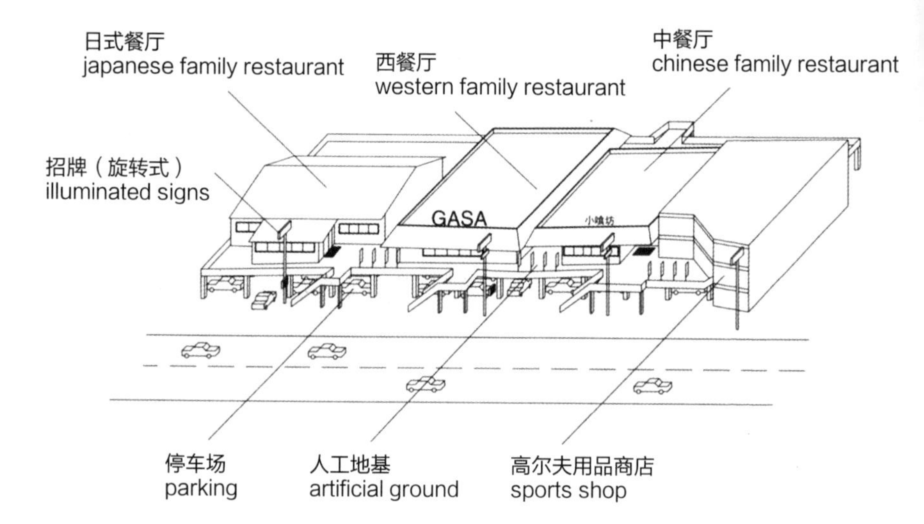

58
家庭餐厅三兄弟
family restaurant triplets

function: family restaurants + sports goods shop + user parking
site: Minaminagasaki, Toshima-ku
- alongside Mejiro road
- 3 restaurant buildings and a sports shop building line up on top of artificial ground
- the ground level parking under the artificial ground and beyond
is shared with the golf practice range on the site behind
- make your choice from Japanese, Western or Chinese cuisine

功能：果蔬市场+店铺

位置：新宿区北新宿

果蔬市场的楼顶被改造为停车场○周围聚集了经营纸箱、软垫等包装材料的店铺○支撑停车场层的柱子兼有通风功能，消除了堆积在下层的蔬菜的异味○道路时而折叠，时而展开，形成了流动的空间

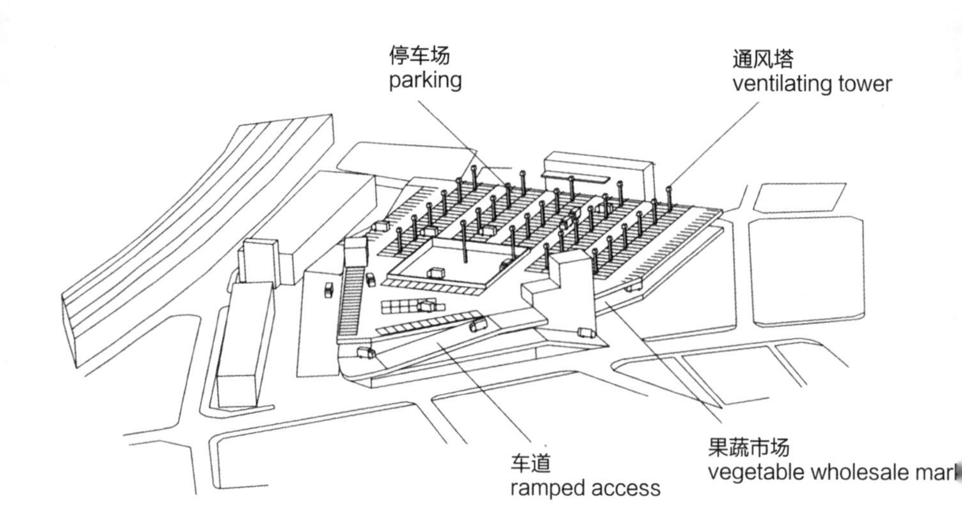

59

果蔬小镇

vegetable town

function: vegetable wholesale market + related shops
site: Kitashinjuku, Shinjuku-ku
- the multi level market can be accessed from many different levels and directions
- the roof terrace of the building is market car parking,
and location of loading and initial exchange
- the surrounding area is filled with shops for packaging, boxes and stuffing
- structural columns double as ventilation pipes,
to take away the smells of raw vegetables
- fluid space obtained by the folding and enlargement of the road

功能：车站+超市+店铺

位置：目黑区大冈山+大田区北千束

东急目黑线（大井町线）大冈山车站○因为改造后的车站位于地下，地上的空地就盖起了轻型建筑物○在原先的铁路（松树根）上开了一层私营铁路经营的超市和商店（松茸）○类似的大楼正在东急线沿线快速增殖

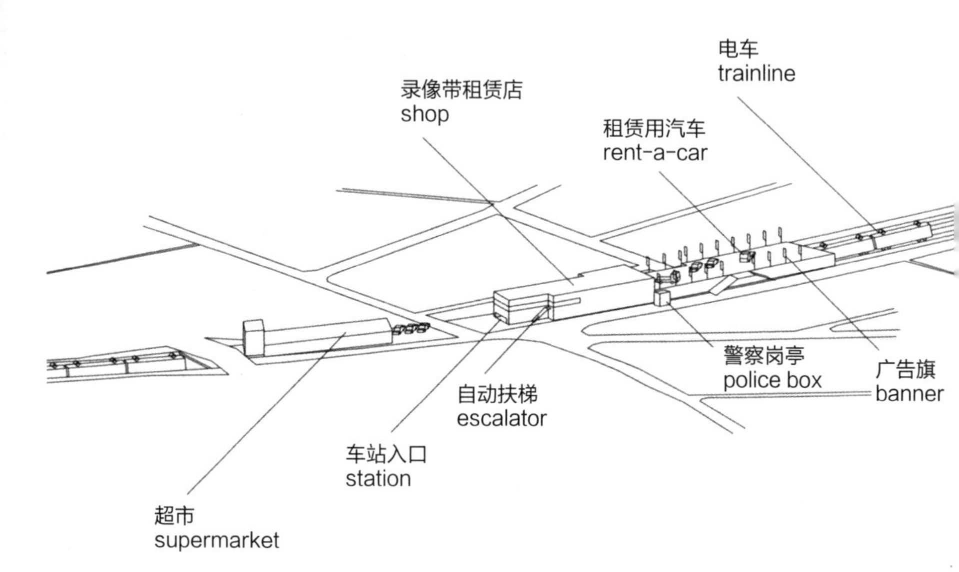

60

松茸形车站楼

sprouting building

function: train station + supermarket + shops + rent-a-car
site: Ookayama, Meguro-ku + Kitasenzoku, Ota-ku
- rail interchange station of Tokyu Meguro and Oimachi lines
- relocating the station underground makes a vacancy
on the ground floor quickly filled with lightweight building
- the railway line (root) company 'Tokyu' develops
the company owned supermarket and shops (sprouts) above
- similar buildings are sprouting all over the Tokyu railway system

TSUTAYA
本
TSUTAYA.
BOOK
VIDEO
CD
GAME
DVD
TSUTAYA
TOKS

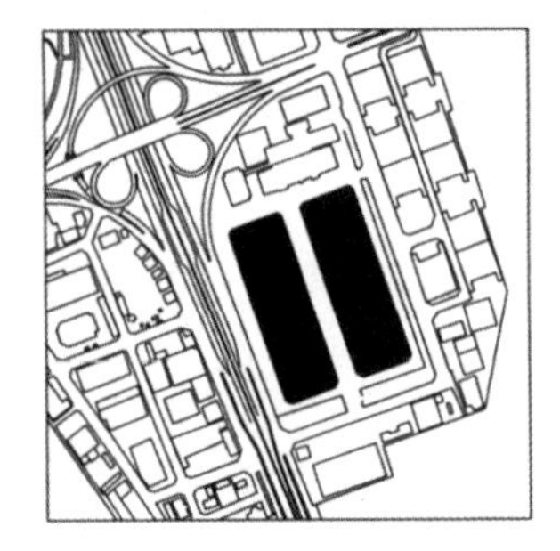

功能：仓库+货运站+停车场

位置：大田区平和岛

位于首都高速公路平和岛出入口前，长300米的巨大双子楼○建筑中央是仓库和货运电梯，四周则是一圈圈车道○斜坡使建筑的外立面充满动态，会让人误以为有沉降差○也许立体停车场是如今流行的现代建筑的鼻祖？

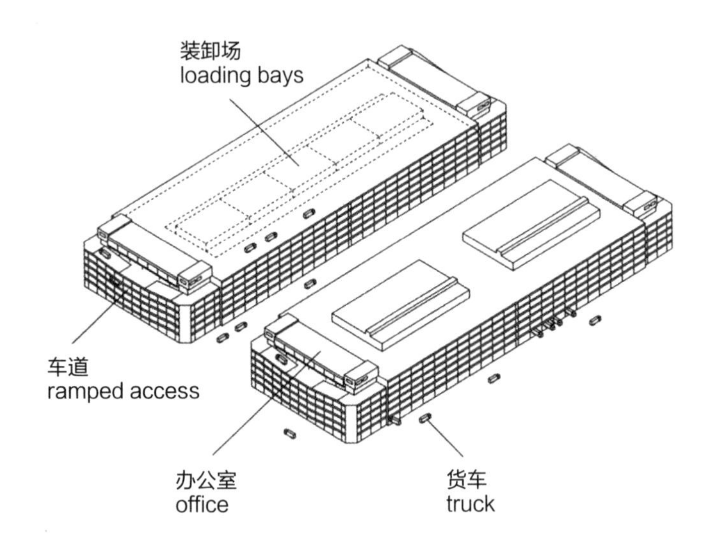

61

TRC（东京物流中心）

tokyo dispersal centre

function: storage station + truck parking
site: Heiwa-jima, Ota-ku
- huge twin buildings each 300 metres long,
stand in front of the Heiwa-jima exit of the expressway
- the centre of each building is filled with storage and elevators,
the rim is ringed with sloping truck circulation
- the end facades are dynamic due to the visible diagonal
of the ramp, and look like differential settlement
- perhaps multi-storey parking is the ancestor of contemporary architecture?

B·3

功能：冷冻仓库+办公室+停车场
位置：大田区平和岛
九个并排而立的大型冷冻仓库位于京滨运河和东京单轨电车之间○卸货平台和办公室把停车场围了起来，背后是巨大的冷冻仓库○厚厚的混凝土墙上开出一扇扇锥形窗户，颇像瑞士山区民居的风格○冷却塔的声音就是冷冻园区的背景音乐○这里就像是城市的大冰箱

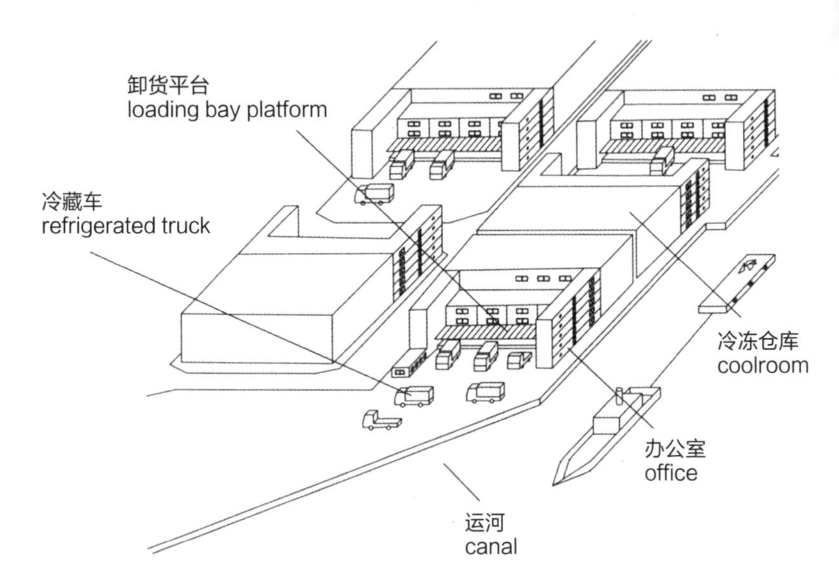

62

TRD（东京冷冻园区）冷冻园区建筑群

coolroom estate

function: coolroom storage units + office + truck parking
site: Heiwa-jima, Ota-ku
- sandwiched between the Keihin canal and
the monorail is a line of nine enormous coolroom units
- offices and the truck loading platform enclose a parking area
- the thick concrete walls of the coolroom are punctured with openings tapering
to small windows in the style of the traditional Swiss cottage
- background music of the refrigeration radiator units
- refrigerator for the city

功能：自动贩卖机+办公室

位置：品川区大崎

狭窄的空地上自然而然地建起了小型建筑物○自动贩卖机犹如房子的武装○类似的例子不可胜数

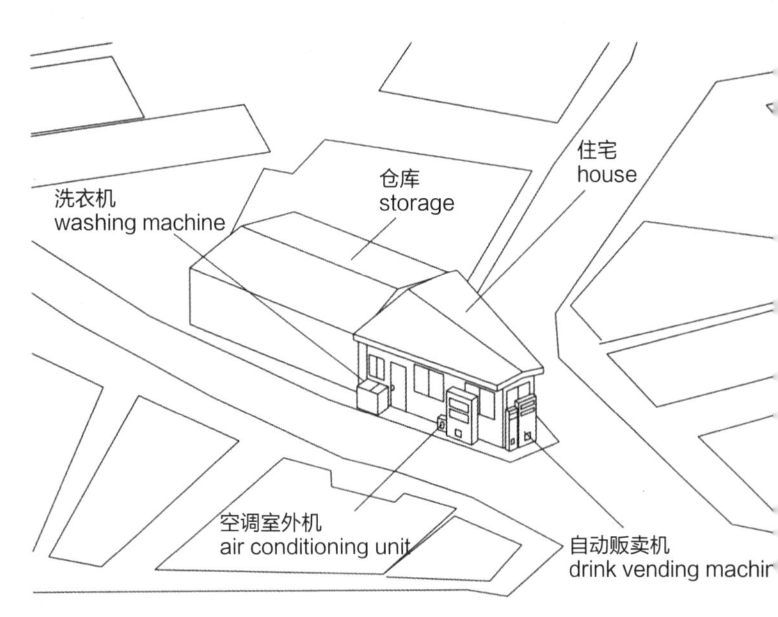

63

宠物建筑 1 号

pet architecture 001

function: house + drink machine

site: Osaki, Shinagawa-ku

- a single storey house on a tiny triangular site
- a small landmark within an industrial area
- armed with drink machines to two sides and the tip
- there are many other examples of this phenomenon

UCC
UCC
UCC
COFFEE
チャオ
Asahi

功能：调节池+公园+公寓

位置：中野区松丘、新宿区西落合

位于妙正寺川调节池上的集合住宅○建筑学与河工学的结晶○建筑外的地面平时是公园，涨水时会被水淹没○这种城市居住形式对于自然的降雨模式十分敏感○建有一块可以对墙练习的网球场

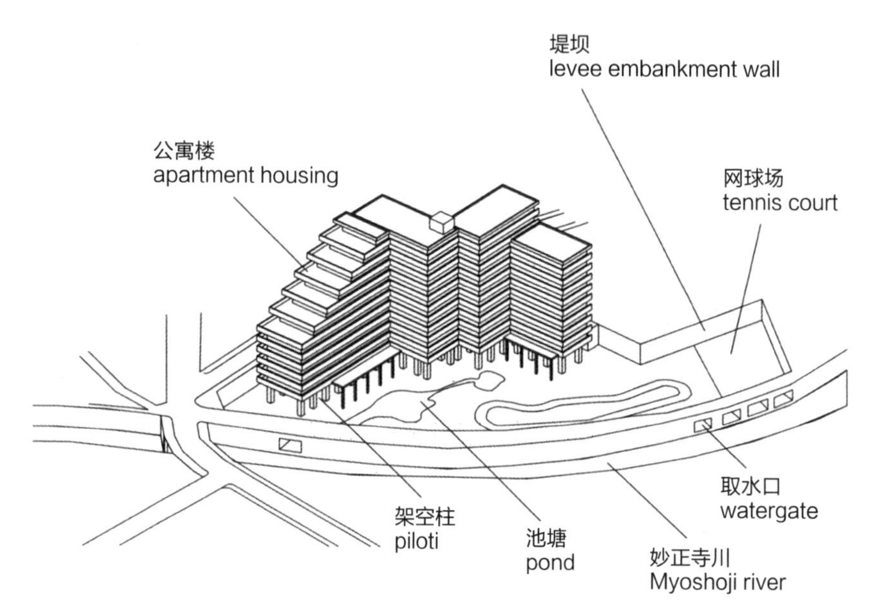

64

水坝公寓

dam housing

function: reservoir + park + apartments

site: Matsugaoka, Nakano-ku, also Nishi-ochiai, Shinjuku-ku

- apartment buildings over reservoir for Myoshoji river
- the base of the apartment buildings are a park at normal times, automatically converting into a reservoir at times of heavy rainfall
- urban living can still retain a high consciousness of natural rainfall patterns
- the tennis wall can be converted to a new game of water tennis
- a confluence of architecture and river engineering

功能：机场航站楼

位置：中央区日本桥蛎壳町、箱崎町

东京的机场航站楼○位于滨町匝道旁边，首都高速公路6号线和9号线交会处下方，长度足有400米○首都高速公路的支路从航站楼中穿过○在某种意义上，这座航站楼是隐身的，几乎看不到它的建筑主体

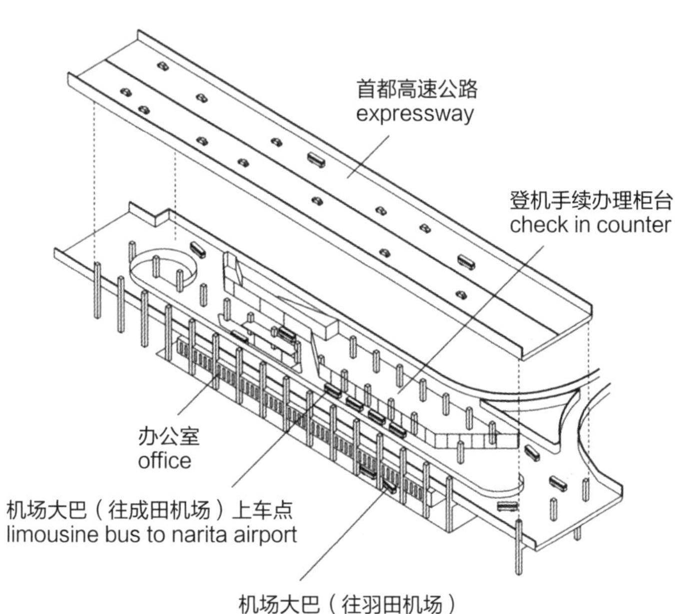

65

立交枢纽航站楼

airport junction

function: airport terminal
site: Hakozaki-cho, Chuo-ku
- inner city airport terminal
- 400 metre long building under the junction of expressway no.6 and no.9 and beside the Hamacho exit
- sections of the main expressway route and branches penetrate this building at mid level
- a kind of ‘stealth’ architecture, whose volume is almost imperceivable

羽田空港へ25分
羽田空港行バ
Bus for Hane

功能：网球场+首都高速公路隧道

位置：涩谷区涩谷

位于青山学院网球场的正中央○像是浮在首都高速公路和六本木大道上方的网球场○到了晚上，网球场周围的挡网会发出亮光○来这里运动的人可以体验在城市中心打网球的乐趣○只是需要留心汽车尾气

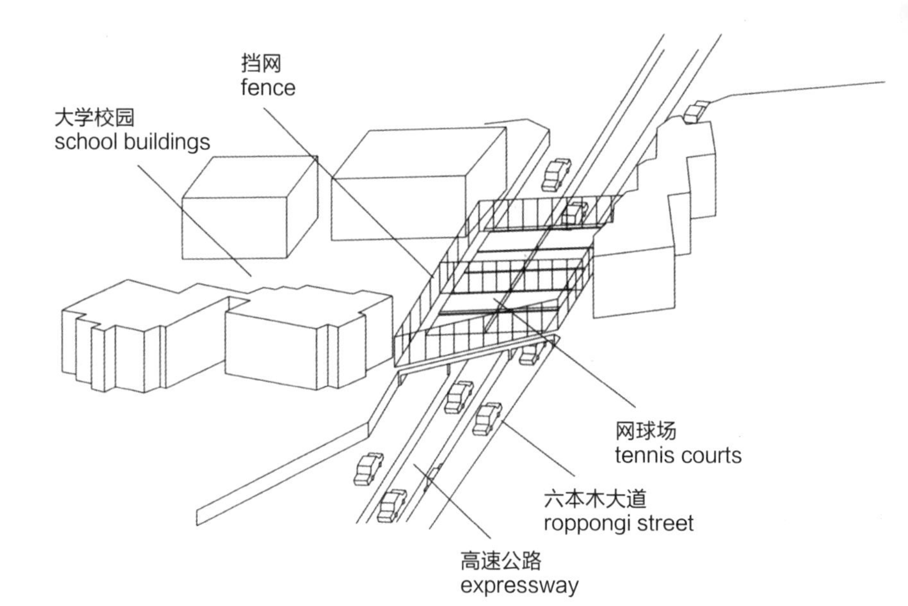

66

运动桥

sports bridge

function: tennis court + expressway tunnel
site: Shibuya, Shibuya-ku
- courts on the rooftop of Aoyama tunnel,
over Roppongi street and the metropolitan expressway
- tennis courts on the premises of Aoyama Gakuin private school
- the courts are next to the sports ground,
which is central to the primary, middle and high schools
- possibility of playing tennis whilst watching the stream of cars below

功能：高尔夫球练习场+网球场+俱乐部会所+集合住宅

位置：目黑区上目黑

一系列被插入树林间的运动“笼子”○高尔夫球练习场利用天然坡道回收球○从斜坡下方向上击打，打出去的球就会沿着斜坡自动滚回○俱乐部会所里不仅有咖啡店和更衣室，还有一个室内高尔夫球场○人类在“笼子”里做各种各样的运动，这里好像一个“人科动物园”

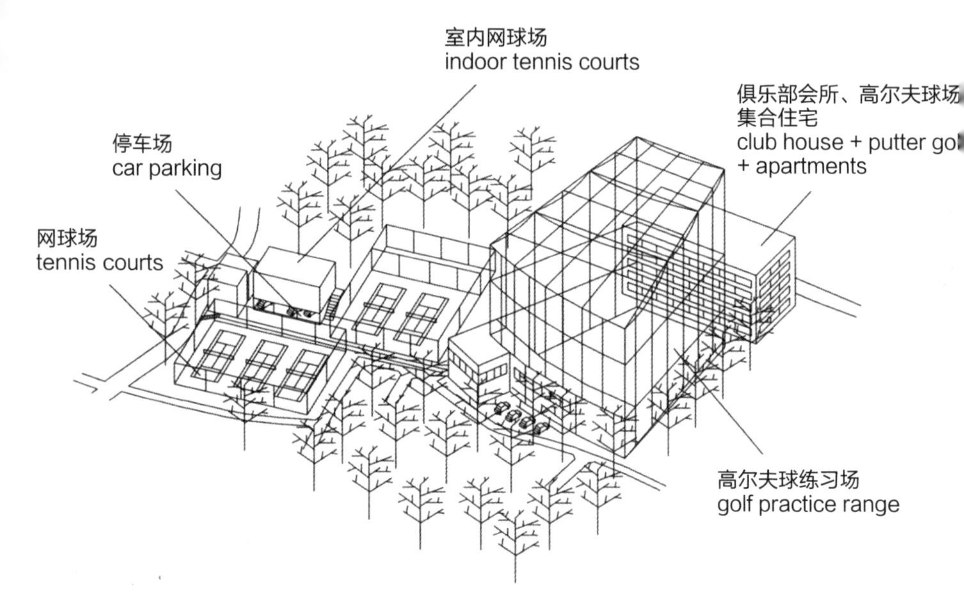

67

运动者动物园

sportsman zoo

function: golf practice range + tennis courts + club house
site: Kami-meguro, Meguro-ku
- a gathering of sports cages in the woods
- the golf range utilises the slope of the land for the ball return
- the club house does not only include a cafe
and changing rooms, but also an indoor putter golf course
- the vision of humans playing various sports
in these cages is just like going to the zoo

功能：仓库+办公楼+直升机停机坪+广告牌
位置：港区海岸
位于港湾大道和运河间一块狭长的土地上，高速公路和东京单轨铁道就在附近○从海陆空三路都可以到达的仓库通过停机坪和办公楼的楼顶相连○停机坪的底座被改造成广告牌○每隔30分钟就有一架直升机起飞，每周日停飞○周边有众多东京湾岸地区的标志性景点，例如彩虹大桥和台场

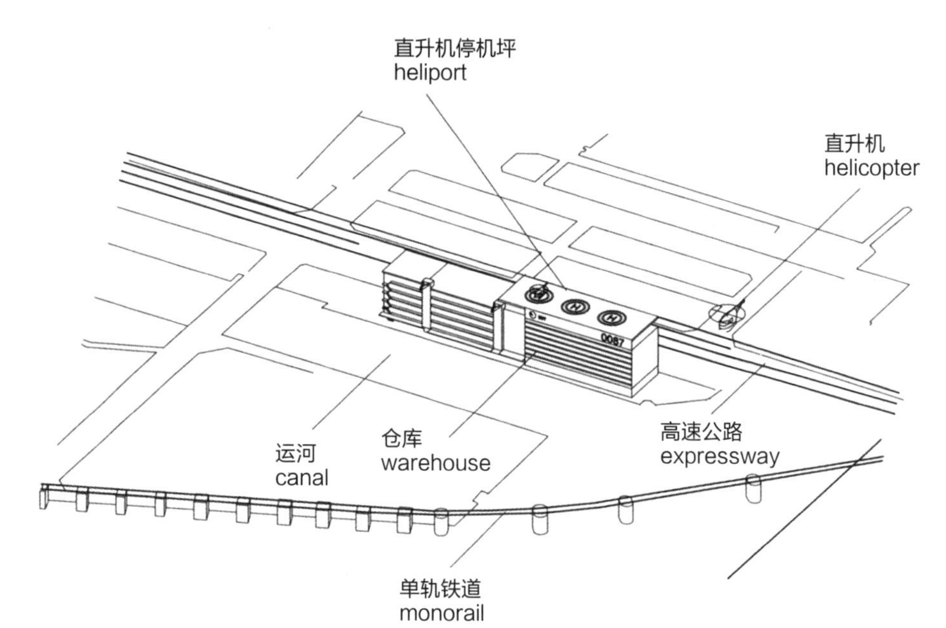

68

直升机仓库

heli-warehouse

function: heliport + warehouse + billboard
site: Kaigan, Minato-ku
- a narrow site sandwiched between the bayside road and a canal with an expressway to the side
- the warehouse can be accessed from land, sea and sky
- the crowning ring of billboards becomes the podium for the heliport
- helicopters come over every half hour, interacting with the bayside symbols of rainbow bridge, ferry terminal, and Odaiba

DDI
DION
国土総合建設

功能：洗车场

位置：涩谷区代代木

山手大道扩建的副产品，建在东京都计划道路区域内的临时设施○用金属材料搭建，机械感十足，外观像是一件大型玩具○二层的手动洗车区和一层的自动洗车区都设有单独的入口，顾客可以自由选择○自动洗车采取的是可以直接驾车通过的“免下车”形式○机器和人类在不同楼层默默竞争，看谁的洗车技术更加高超

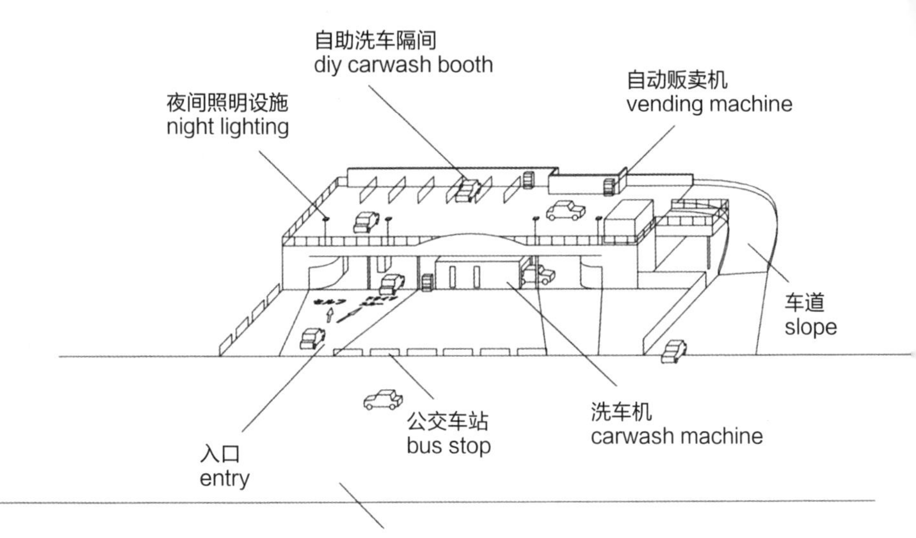

69

洗车露台

carwash terrace

function: diy carwash + machine carwash
site: Yoyogi, Shibuya-ku
- temporary facility utilising land set aside
for future widening of Yamate street
- metallic and machinistic, yet toy like appearance
- the machine carwash is drive through
- at the entry point, drivers select from upper level
diy carwash and ground level machine wash
- can your skills and speed compete with the machine?

セルフ
ドライブ
スルー

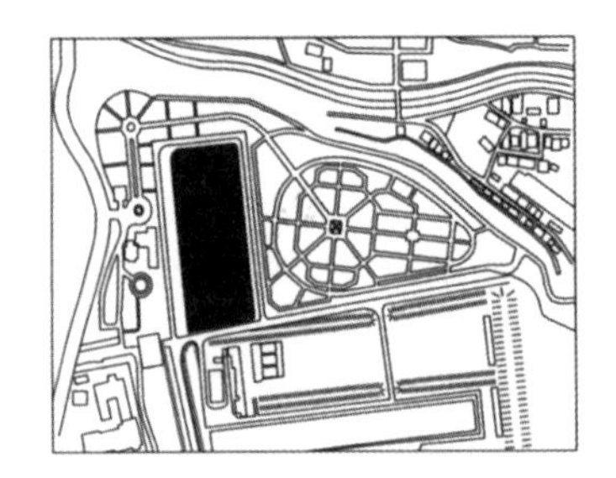

功能：射击练习场+墓地

位置：新座市新塚

1964年东京奥运会射击比赛的场地，利用了墓地的地形○练习场周围被土堤环绕，现在供自卫队队员进行射击练习○从墓地虽然无法看到自卫队队员，但可以听到从练习场传来的枪声○枪声和墓地连成一片，想来不禁让人脊背发凉

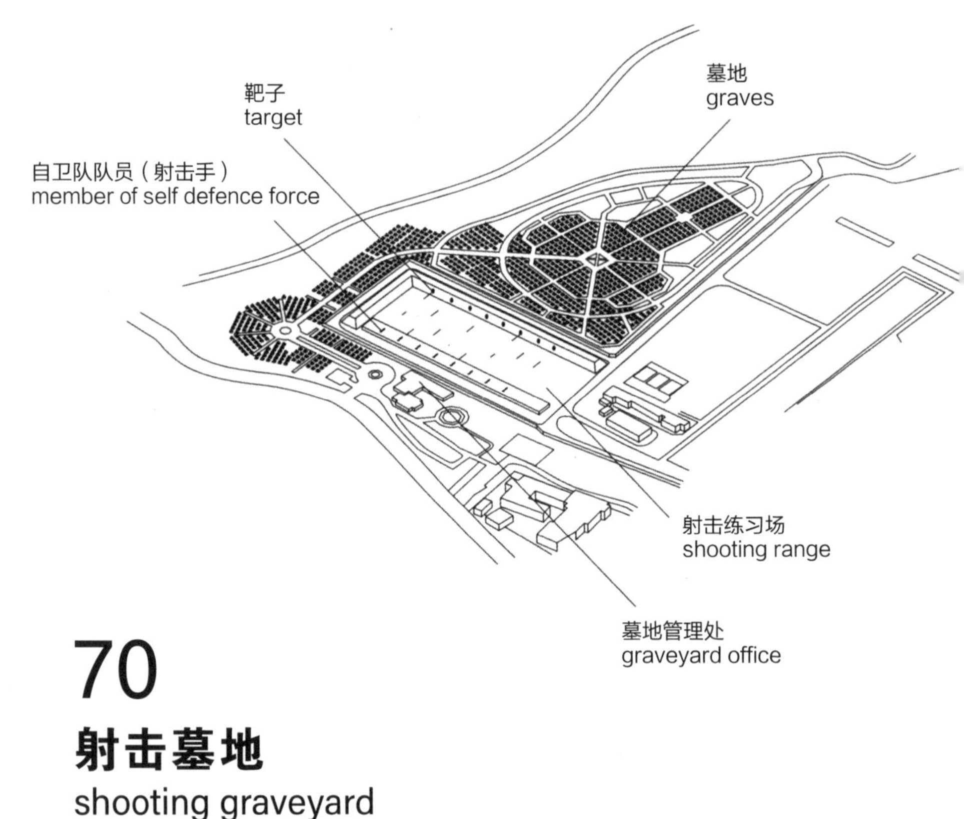

70

射击墓地

shooting graveyard

function: shooting range + graveyard
site: Niizuka, Niiza-shi
- site for the Tokyo Olympic shooting range,
utilising the grounds of an ancient tomb
- now it is used for practice by the self defence forces
- it is hidden by surrounding embankments,
but the sound of shooting is very near
- the adjacency of guns and graveyard: a chilling thought

MAP
地图

1 仓库球场
warehouse court

2 电器走廊
electric passage

3 高速百货商店
highway department store

4 电影桥
cine-bridge

5 过山车楼
roller coaster building

6 霓虹灯楼
neon building

7 柏青哥教堂
pachinko cathedral

8 风俗楼
sex building

9 卡拉OK宾馆
karaoke hotel

10 首都高速公路巡逻站
expressway patrol building

11 大使馆楼
embassies building

12 公园停车场
park on park

13 公交住宅区
bus housing

14 高尔夫出租车站
golf taxi building

15 混凝土公寓
nama-con apartment house

16 汽车塔
car tower

17 马公寓
horse apartment house

18 物流综合体
distribution complex

19 空调楼
air-con building

20 广告公寓
billboard apartment house

21 神社楼
shrine building

22 渣土公寓
sand apartment house

23 配送螺旋
delivery spiral

24 澡堂旅游楼
bath tour building

25 出租车楼
taxi building

26 货车塔
truck tower

27 立交球场
interchange court

28 跃层式加油站
double layer petrol station

29 超级驾校
super car school

30 污水球场
sewage courts

31 净水球场
supply water courts

32 墓地通道
graveyard tunnel

33 阿美横町空中寺院
ameyoko flying temple

34 商店崖
shopping wall/mall

35 铁道博物馆
rail museum

36 双子污水花园
twin deluxe sewerage gardens

37 增殖滑道楼
proliferating water slides

38 通风方尖碑
ventilator obelisk

39 车站上的家
apartment station

40 蜈蚣住宅
centipede housing

41 汽车村
vehicular village

42 潜水塔
diving tower

43 底盘公寓
chassis apartments

44 TTT（乐高办公楼）
TTT (lego office)

45 隧道神社
tunnel shrine

46 公寓山寺
apartment mountain temple

47 吸血公园
vampire park

48 起重机架
crane shelves

49 幽灵铁道工厂
ghost rail factory

50 挡土墙公寓
retaining wall apartments

51 桥洞里的家
bridge home

52 宅地农场
residential farm

53 物流立交枢纽
dispersal terminal

54 皇家高尔夫公寓
royal golf apartments

55 汽车办公楼
car parking office

56 绿色停车场
green parking

57 汽车百货商店
auto department store

58 家庭餐厅三兄弟
family restaurant triplets

59 果蔬小镇
vegetable town

60 松茸形车站楼
sprouting building

61 TRC（东京物流中心）
tokyo dispersal centre

62 冷冻园区建筑群
coolroom estate

63 宠物建筑1号
pet architecture 001

64 水坝公寓
dam housing

65 立交枢纽航站楼
airport junction

66 运动桥
sports bridge

67 运动者动物园
sportsman zoo

68 直升机仓库
heli-warehouse

69 洗车露台
carwash terrace

70 射击墓地
shooting graveyard

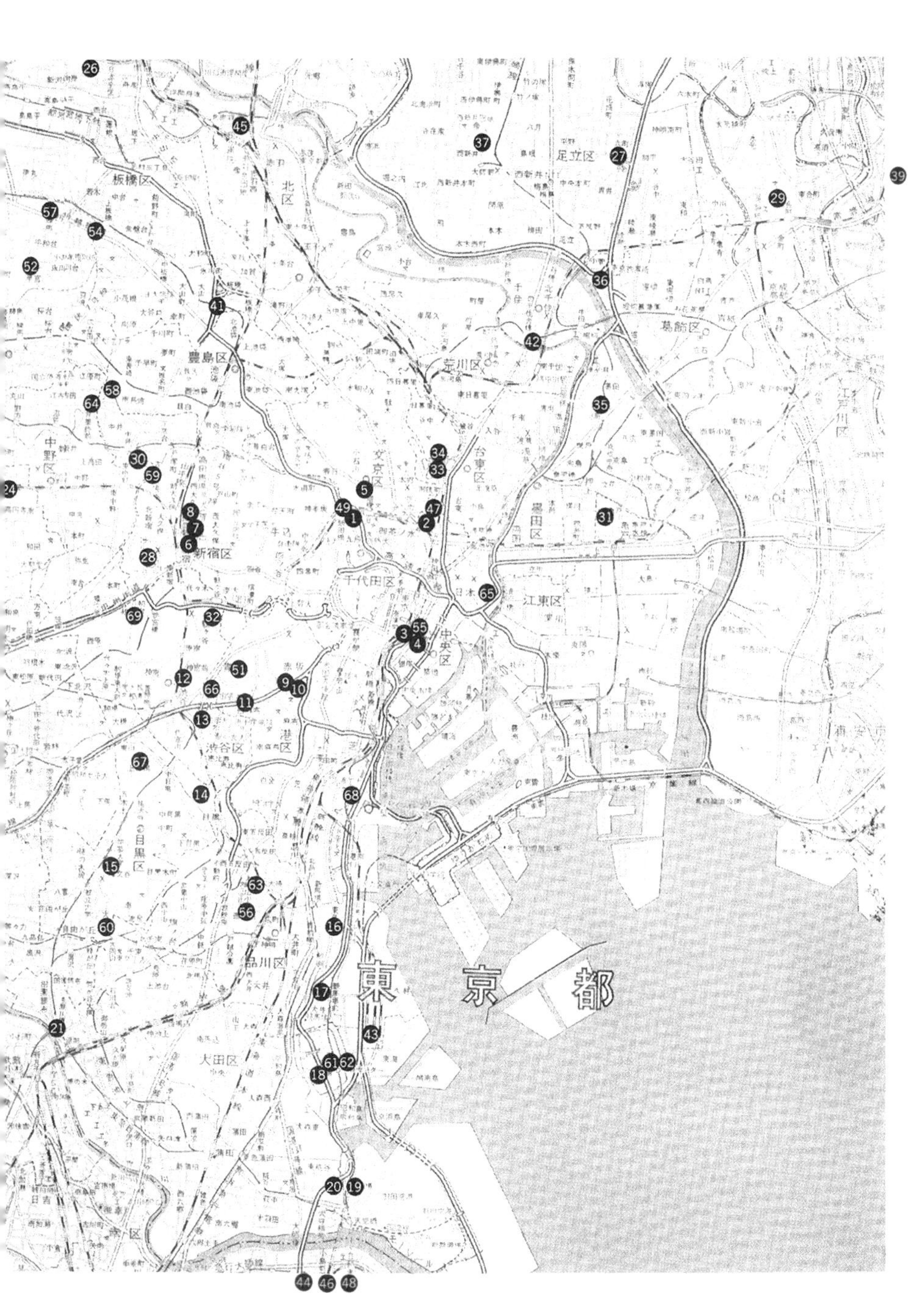
板橋区
北区
足立区
豊島区
荒川区
葛飾区
文京区
台東区
中野区
新宿区
墨田区
千代田区
江東区
中央区
渋谷区
港区
目黒区
品川区
大田区
東京都

TALK WITH TAKASHI HOMMA
对话本间隆

2000年5月7日 本间隆（TH）× 贝岛桃代（MK）+黑田润三（JK）+塚本由晴（YT）
TH: Takashi Homma × MK: Momoyo Kaijima + JK: Junzo Kuroda + YT: Yoshiharu Tsukamoto
May 7, 2000

“哈，真有意思！”

MK：首先我想请教本间先生，您作为建筑师、摄影师在东京工作、生活，对《东京制造》有什么看法呢?

TH：你是不是正式谈论过它和赤濑川源平的超艺术托马森之间的差异来着?

MK：超艺术托马森尝试把漫步街道时捕获的城市记忆的碎片带往现实的城市空间，讲述无数微小历史的故事。所以，其中总有种怀旧的静谧氛围，只是回首往昔，难以展望未来。因此，当我们谈论东京实际上正在建造什么样的建筑物时，它就很难说出什么了。我认为了解我们能如何使用城市是重要的，所以想记录下人们对于未来城市的想法和愿景。

TH：原来如此，并不是怀旧呀。那么，既然“东京制造”的建筑物总被认为是“失格建筑”，您为什么对它们那么感兴趣呢?

MK：从所谓的建筑作品的常识看，这些建筑物的确都是些“失格建筑”，不过仔细观察就会发现它们都很有意思。所以我想，“失格”不意味着一种否定，而是一种让人喜爱的东西。

YT：也有那种因为太过严肃而和现实世界显得格格不入的建筑，我觉得它们的严肃里有种幽默感，好像给这个大城市按下了 OFF 键似的。

TH：东京是个经常让人觉得紧张的城市，“东京制造”的建筑让我看着看着就觉得“哈，真有意思！”，感觉轻松了许多。

MK：我们就希望读者有这样的感受。在类似东京这样高密度的混乱城市里，想要严肃认真地发起挑战可是件费力的事啊。(笑)

TH：我认为“哈，真有意思！”这种力量挺重要的。那个水泥工厂（“混凝土公寓”）就荒谬得很可爱。“电器走廊”之类的名字也妙不可言。

平面

TH：我看到除了实景照片之外，你们还绘制了线条图。是想象着建筑物的样子画的吗?

JK：其中加入了我们自己的理解。虽然图是大家分头画的，但因为是电脑绘图，所以没有表现出各自的风格。后期制作时也没怎么通过渲染细节和表现光线来让手绘的建筑看起

TH: The buildings of Made in Tokyo are often described as ‘no good’ (da-me). Why is it that you started to become interested in this quality?
MK: These buildings are regarded as da-me from the point of view of architectural magazines, but if you look closely at them they are very interesting. So, this da-me is not only negative, but also adorable.
YT: They are too serious – in fact many of them have become a caricature of their urban reality. Their da-me includes humour pouring out of their ridiculous seriousness, and release from the city.
TH: Made in Tokyo seems to let me laugh freely, while in amongst the tension of Tokyo.
MK: We are after exactly that feeling. After all, trying to pick a fight with the high density and confusion of Tokyo is pretty hard. (laughter)
TH: At the beginning, I thought the drawings were a lot cooler than the photos. If it was only the photos, maps and explanations, it would be quite normal, and actually fairly hard to understand. But when I get to the drawings, it really hits me.
MK: Is that because you are a photographer?
TH: Well, in fact I seem to be able to see more surface plane in the drawings.
YT: When we tried to look at the buildings as flatly and without value judgements as possible, they became black and white line drawings. With a photo, you can’t keep out the nuances of neon colours, urban grime etc.
TH: I think that for an everyday person, it is the drawings that make clear what is so exciting about the photos.

来接近实物，而是通过内部透视，突出了想强调的细节。

TH：我一开始觉得线条图比照片更酷。如果只有照片、地图和解说，那也太普通了，况且还很可能看不明白。不过这些图可真出乎我的意料。（笑）

MK：因为您是摄影师吗？（笑）

TH：该怎么说呢……线条图这种形式，能把事物放在一个平面上。

YT：把一座建筑物完全放在一个平面上，而不考虑它们的意义，我最终完成的就是黑白的线条图。如果是照片的话，不免就会把霓虹灯的颜色啦，表面的脏污啦之类的细节拍下来。

TH：没错。我拍"高尔夫出租车站"的时候也觉得照片里会闯入不少多余的东西，真让人伤脑筋。画下来的话，多余的东西就可以通通去掉。

傻画

MK：而且我们会尽量画得可爱一些，把汽车以及建筑物周围的东西都画上也很重要。因为这样能表现建筑物和周围环境的关系。

JK：把搅拌车的细节也完整画出来的话，效果就会完全不一样。马也画得比一般的马要大一些，可不是按着赛马的比例来画的。（笑）

TH：建筑物的比例是对的吗？

JK：比例做了调整。大型建筑物会变扁，所以进一步做了变形处理。

YT：发现建筑物，拍下照片，下笔作画。画这些图也很费功夫，需要经过训练，下笔时才有感觉。如果每根线都画得一样粗，就表现不出情感了。虽然我们画的是每个建筑物的远景，实际上每幅画里都包含着我们的喜爱之情。

TH：如果把照片拿给一般人看，跟大家解释说"这是公交住宅区，很厉害吧！"，应该没人会懂吧。但画就不同了，能让人很快明白"啊，是那样啊"。也就是说，画能告诉人们应该如何去观察。

MK：藤森先生他们则想用文字告诉人们应该如何观看。

JK：历史学家们也让我们多用文字。

TH：千万别，把这些写成故事讲出来可就太糟了。

MK：我们只是如实地描写了眼前看到的东西，为这些建筑物预留了未来的空间。所以，看起来就是这样简单的风格。

YT：只是照片的话，难免偶尔会带上批判的眼光。有一种"这样的建筑物真的没问题吗？"的感觉。

TH：搞得好像引起了什么社会问题似的，某本周刊杂志上经常有这样的报道吧。

MK：也有人把"东京制造"看作是对城市现状的某种批判……本间先生，对您的《东京郊外》的评论里也有这种声音吧？

TH：事实上，有很多人认为《东京郊外》是在批判城市现状。

MK: It is what Fujimori-san was trying to do in words. (Terunobu Fujimori et al, Institute of Street Observation)
JK: Often, historians say to us that we should put it more into words.
TH: Oh no! If you turn it into a story, that will be the end .
MK: It has become practical, because we simply concentrated on describing what was in front of us, and we wanted to leave room for the future.
YT: If there are only photos, then they can be taken as just criticism. Like 'is this building really ok?' But I think that by making the drawings, it can't be seen as purely critical any more. These drawings are foolish, innocent, speedy...
TH: I think the way the work passes right through the criticism is really good. The drawings are hugely important.
YT: It was very memorable for me that when we went together to the nama-con apartment house, you didn't take any photos. You said 'hmm – it just doesn't give joy without the drawings and text.'
TH: Rather than the building itself, the interest is guided by the drawings of 'Made in Tokyo.'
MK: The Tokyo which we see, and the Tokyo which you see, are different aren't they. When I saw your works of 'TOKYO SUBURBIA', on the one hand I thought 'oh, this is entirely different from Made in Tokyo,' but then on the other hand, I thought 'yes, I can understand this.' I think it is interesting that we want to see different aspects of the same Tokyo.
TH: If anything, I hope Tokyo can sustain at least that much. It wouldn't be interesting if just one single

YT：但这些画总不会让人联想到批判吧。它们看起来很傻，或者说很天真，又或者说很纯净。

TH：真好啊。这些画让人感觉豁然开朗，它们真的很重要。

超越批判

YT：本间先生当时也被我带到混凝土公寓去了，但他在现场只是不停若有所思地发出“嗯……”的声音，到最后什么也没拍，我对此印象深刻。“果然没有图画和解说的话，会失去很多乐趣呢。”（笑）

TH：通过《东京制造》这样带有图画和解说的导览手册看到的建筑，比实际的建筑要有趣得多呀。（笑）

MK：我们这些人看到的东京和本间先生看到的东京应该不一样吧。我读了本间先生的《东京郊外》，既觉得和《东京制造》完全不同，又有一种“原来如此”的豁然开朗之感。明明是同一个东京，看起来却像完全不同的两个地方。这点很有意思。

TH：荒木先生和篠山先生也都说过“我拍的是东京”，一些建筑师也这么说。但是东京的面孔太多了，不是一个人能拍得完的。所以，我眼中的东京当然和《东京制造》里的东京不同。《东京郊外》收到了许多负面评价，说“那些地方根本就不能代表郊外！”，其实我也没有说“这些地方就是东京的郊外！”。我们看到的终究只是局部。不过，“东京制造”里有 70 座建筑物，可真厉害呀。而且在里面活动、居住的人什么都不知道，可真有趣呀。

轻型城市论

MK：大部分建筑师应该都像勒·柯布西耶一样更关注作为一个整体的城市吧？这也许是正统的城市论，但不是我们此刻面对的现实。我们不认为某一个个体能看到完整的东京，因为东京实在是太大了。所以，《东京制造》呈现的是一种轻型城市论，虽然不是正统，却也因此才有了具体地捕捉城市的可能性。

TH：东京这座城市就是像这样才好，如果让某一个人都道尽了，反而没什么意思。“东京论”这种说法其实十分傲慢。难得在东京这样一座城市探索，我们试着发现一个更加巨大的东京。如果都被一个人归纳总结完了，那还有什么可探索的，简直荒唐可笑。而且，东京也不是三两下就能被研究透的。

我觉得那些对《东京制造》和《东京郊外》大发牢骚的人，其实只是急着想搞清楚这两本书对东京到底是褒还是贬罢了。

MK：就这样不清不楚的才好嘛。

TH：而且明明就搞不清楚嘛，这事没那么简单。

JK：我第一次办展览的时候，建筑学会事务局的人也对“失格建筑”这个说法感到十分担忧。还说了“要是出了问题怎么办”之类的话。

YT：不过意外的是我们没接到多少投诉。

person was able to completely explain everything. Since this is where we are working, I'd like Tokyo to be gargantuan! If someone explains the whole city, it becomes meaningless and stupid to work here. Anyway, I think Tokyo just isn't such a simple thing. Whether looking at 'Made in Tokyo' or 'TOKYO SUBURBIA,' people who are irritated by our work are ones who want a definite position to be stated – are we being critical or appreciative?
MK: I think it's better to not be definite.
TH: More than anything, I feel that 'Made in Tokyo' is enjoying Tokyo. It is useless to criticise and go all dark and narcissistic, saying that the olden days were so much better. People usually like the past. Even in photography, there is a tendency towards the nostalgia of sepia toned images, and critics often pick up on these. I do think that because everybody likes them, there is a high demand and so they are necessary. But if you aren't going to get appreciated, a different method of expression can be fruitless and tiring. It seems that photography is of the past after all. In terms of fact, it can only take images of the past. People become satisfied with nostalgia. But if that's all, I think photography has no future. I get annoyed by the way that critics don't seem to realise this.
YT: But, isn't it possible to look forwards, depending on what you fix into the image?
MK: In the collection of 'Made in Tokyo,' some of the buildings have already been demolished. Someone

MK：等这本书一出版，抱怨声肯定此起彼伏。

TH：有抱怨不也挺好的嘛。

MK：因为我们是在城市环境里做研究，抱怨越多，不就越能证明我们的研究和这座城市的关系十分紧密吗？

TH：《东京制造》首先给人的感觉是在享受东京吧，不要乱给它扣帽子嘛。（笑）假如有读者带着点暗黑的自恋念叨着以前怎么怎么好，对着它批判一番，是毫无意义的。

拍下东京的未来？

MK：你不是拍过这样的照片吗？老街区的背后冒出新宿的摩天楼。我们看待东京的方式和这样的照片有什么不同呢？

TH：人还是恋旧啊，拍照片到最后也是要回到过去。深棕色的画面本身就充满了怀旧的色彩，评论家们似乎也很买账。大家都喜欢，所以就有需求，这样的照片就必然出现。如果得不到肯定，尝试新的表现手法就只是吃力不讨好……（笑）我觉得照片果然是属于过去的东西，只能留下属于过去的影像。所以大家都觉得留恋过去是一件好事。但只是这样，摄影的未来又在哪里呢？评论家们怎么就没有注意到这一点呢，真让人着急。

YT：不过，如果使用不同的显影方法，还是可能拍出面向未来而非怀念过去的照片吧？

MK：即便是“东京制造”的建筑物，有些实际上也已经不在了。记得有人问过我们，为什么不发起保存旧建筑的运动，这话听起来很不现实。

TH：不要发起什么保存旧建筑的运动，就让它们被拆掉，然后再建起来一栋栋闪闪发亮的新建筑，这样也挺好的。（笑）

MK：如果我们把每座建筑物都看成是东京现状的一份报告的话，那么当它不再能发挥这一功能时，随之消失倒也无妨。这么看，这些建筑物都有积极向前、拼命地活在当下的感觉，真有意思啊。

YT：我觉得如果能把这种感觉定格在照片里，那么摄影也就有了未来。

TH：但有这种感觉的人毕竟是少数吧。一边想着做这些事不被人理解，一边还是继续做着。我和塚本先生身上总带着这样的自虐倾向。不过贝岛先生倒是认为这些狭小的城市剖面总有一天能被大众看到和理解。贝岛先生的想法太有意思了，这也是我们大家共同的愿望。即便我说了不算数，贝岛先生一说，大家就欣然附和了。（笑）

MK：……话说回来，你觉得这书卖得出去吗？

TH：这谁知道呢。

MK：你的意思是卖不出去吗？

TH：我的意思是卖出去可就太好了。（笑）

asked me 'why don't you start up a preservation society?', but I think there is no reality in this.
TH: Isn't it better to demolish and rebuild a shiny-new, much more fantastic version? (laughter)
MK: If we understand each building as a measure of the current condition of Tokyo, surely it's ok for it to disappear once this role has become redundant. I am interested in the forward looking quality, the throw away feeling, in any case really-living-right-now sense of these buildings.
YT: I feel that there is a future in capturing those feelings through photography.
TH: But I think this way of thinking is in a minority. I keep doing it, even though I'm not sure if it is understood. But I think that Kaijima-san believes that everyone can understand. That aspect of Kaijima-san is very delightful, and I think it gives hope to the rest of us.
MK: ...Mmm – anyway, do you think this book will sell? (laughter)

HISTORY
沿革

1991

贝岛桃代和塚本由晴在涉谷发现了“东京制造”传奇的第一号建筑（棒球练习场 + 意大利面餐厅，数年后被拆除）。在这之后的几年，两人又陆续发现了“高尔夫出租车站”“混凝土公寓”等建筑物。
The first “Made in Tokyo” building discovered in Shibuya by Kaijima + Tsukamoto. Several more candidates found over the following years.

1996

“东京制造”作为现代日本建筑装置的一部分，受邀参加“1996 年年度建筑：革命的建筑博物馆”展览（矶崎新策展）。
“Made in Tokyo” invited to take part as the contemporary history component of the exhibition “Architecture of the year 1996: architectural Museum of Revolution”.

05 与黑田润三、藤冈务等人共同成立“东京制造团队”。
T.M.I.T. established with Kuroda and Fujioka.

09 贝岛桃代和策展人矶崎新在佛罗伦萨举办的展览上进行了关于“东京制造”的对话。
[本次展览的手册《矶崎新的革命游戏》（TOTO 出版）里收录了对“东京制造”30 座建筑物的介绍。]
Discussion on the run in Florence, Kaijima and Arata Isozaki about the publication of Made in Tokyo for the exhibition catalogue.

11 作为展览上的一件装置，制作和分发了印有“东京制造”建筑物的 300 件 T 恤。
Exhibition installation including 300 T-shirts.

1997

03,04 《10+1》第 8 期和《InterCommunication》第 20 期介绍了“东京制造”。
Publication in “10+1” vol.8 and “InterCommunication”, vol. 20.

04 塚本由晴在东京工业大学的研讨班上继续推进“东京制造”。
Development by the master course student class in Tokyo Institute of Technology.

07 贝岛桃代在《Space Design》7 月号上发表了《香港制造》。
Made in Hongkong is published in “Space Design” 1997 July. Kaijima Institute Publishing.

09 “东京制造”首次海外个展在瑞士苏黎世的建筑论坛举办。
Exhibition in Architektur Forum Zurich.

1998

06 “东京制造”日文网站完成。
http://www.dnp.co.jp/museum/nmp/madeintokyo/mit.html
Japanese version web site was opened.

12 参加本间隆的“东京郊外”展览：“东京制造的新郊外”（于 PARCO 美术馆）。
Model “New Suburbia Made in Tokyo”. Collaboration with Takashi Homma for the exhibition at Parco Gallery.

1999

02 贝岛桃代和森川嘉一郎、后藤武在第 11 届“现代、城市与建筑”研究会上就“东京制造”展开对谈（于东京大学）。
Panel discussion between Kaijima and Kaichiro Morikawa, Takeshi Goto in Studies of “Modernity/ Urbanism/Architecture.”

05 为 21 世纪学生会议制作《东京回收指南》。
Tokyo Recycle Guide Book in 21st Century Students Conference.

11 “东京制造”英文网站完成，受邀参加威尼斯国际建筑双年展的线上展览。
http://www.dnp.co.jp/museum/nmp/madeintokyo_e/mit.html
English version web site was opened. Invitation to “Expo on line” of Venice Biennale.

2001

02 受邀前往巴塞罗那举办研讨会、演讲和展览。
Workshop, lecture and exhibition in limits col.legi. d'Arquitectes de Catalunya, Barcelona.

PROFILE
作者简介

贝岛桃代

1969 年出生于东京。1991 年毕业于日本女子大学。1992 年与塚本由晴一同创立犬吠工作室（Atelier Bow-Wow）。1994 年完成东京工业大学硕士课程。1996 年获苏黎世联邦理工学院奖学金。2000 年东京工业大学博士课程期满退学。2000 年起担任筑波大学讲师；2009 年起担任同校副教授；2017 年起担任苏黎世联邦理工学院建筑行为学教授。曾在哈佛大学设计研究生院（2003、2016）、苏黎世联邦理工学院（2005—2007）、丹麦皇家艺术学院（2011—2012）、莱斯大学（2014—2015）、代尔夫特理工大学（2015—2016）、哥伦比亚大学（2017）讲学。2017 年起担任 CHEER ART 副理事长。2018 年担任第 16 届威尼斯国际建筑双年展日本馆策展人。主要建筑设计作品都与塚本由晴合作：House & Atelier Bow-Wow（2005）、宫下公园（2011）、桃浦村（2017）、尾道车站（2019）等。主要著作有《建筑民族志》（2018，TOTO 出版，与他人合著）等。

黑田润三

1968 年出生于茨城。1992 年毕业于武藏野美术大学造型学部建筑系。1992 年至 1996 年在东京工业大学坂本一成研究室学习。1996 年创立黑田工作室，开始个人美术、建筑活动。2007 年工作室改组，更名为黑田润三工作室，主要从事建筑设计（居住空间、医院建筑等）、美术指导、企业形象设计等。主要建筑设计作品有七色妇产科诊所、玫瑰泉妇产科诊所、七色小镇、筑波木场公园诊所等。

塚本由晴

1965 年出生于神奈川。1987 年毕业于东京工业大学工学部建筑系。1987 年至 1988 年以旁听生身份在巴黎第八大学学习。1992 年与贝岛桃代成立犬吠工作室。1994 年完成东京工业大学博士课程，获工学博士学位，现任该校教授。2012 年当选英国皇家建筑师学会（RIBA）国际会员。曾在哈佛大学研究生院（2003、2007、2016）、加州大学洛杉矶分校（2007—2008）、丹麦皇家艺术学院（2011—2012）、莱斯大学（2014—2015）、代尔夫特理工大学（2015—2016）、哥伦比亚大学（2017）讲学。2020 年起担任一般社团法人“小小地球”理事。主要设计作品参考贝岛桃代的介绍。主要著作有《图解犬吠工作室》（2007，TOTO 出版，与他人合著）、《世界之窗：窗边行为学》（2010，Film Art 出版社，与他人合著）。

Momoyo Kaijima

1969 Born in Tokyo. 1991 Graduated from Department of Housing, Faculty of Home Economics, Japan Women's University. 1992 Established Atelier Bow-Wow with Yoshiharu Tsukamoto. 1994 Completed the Master Course of Architecture, Tokyo Institute of Technology. 1996-97 Studied at the Federal Institute of Technology (ETH), Zurich. 2000 Completed the Doctor Course of Tokyo Institute of Technology. 2000-2009 An assistant professor at the Art and Design School of the University of Tsukuba, and since 2009, Has been teaching there as an associate professor. Since 2017, Has been serving as a Professor of Architectural Behaviorology at ETHZ. As a visiting professor at the Department of Architecture at Harvard GSD (2003, 2016), a guest professor at ETHZ (2005-07), as well as at the Royal Danish Academy of Fine Arts (2011-12), Rice University (2014-15), Delft University of Technology (2015-16), and Columbia University (2017). Since 2017, Vice Director of Specified NPO, Cheer Art. She was the curator of Japan Pavilion at the 16th International Architecture Exhibition-La Biennale di Venezia. Main works: House & Atelier Bow-Wow (2005), Miyashita Park (2011), Momonoura Village (2017), Onomichi Station (2019) etc. Main books: "Atelier Bow-Wow Graphic Anatomy" (2007, TOTO Publishing), "Architectural Ethnography" (2018, TOTO Publishing) etc.

Junzo Kuroda

1968 Born in Ibaraki Prefecture. 1992 Graduated from Department of Architecture at Musashino Art University. 1992-96 Studied as a research student at the Sakamoto Laboratory of Tokyo Institute of Technology. 1996 Established Atelier Kuroda and started his career as an architect and artist. 2007 Reorganized the studio as Junzo Kuroda Atelier. Business: Architectural design (houses, clinics, and all sizes), art direction, CI design, etc. Main works: "Nanairo Clinic for Women", "Bara-No-Izumi Women's Clinic", "Nanairo Comachi", "Tsukuba Kiba Park Clinic", etc.

Yoshiharu Tsukamoto

1965 Born in Kanagawa Prefecture. 1987 Graduated from Department of Architecture and Building Engineering, Faculty of Engineering, Tokyo Institute of Technology. 1987-88 Studied at Ecole d'Architecture de Paris-Belleville. 1992 Established Atelier Bow-Wow with Momoyo Kaijima. 1994 Completed the Doctor Course, Tokyo Institute of Technology. Currently professor at Department of Architecture and Building Engineering, Faculty of Engineering, Tokyo Institute of Technology. Doctor of Engineering. Taught as a visiting professor at the Department of Architecture at Harvard GSD (2003, 2007, 2016), UCLA (2007-08), as well as at the Royal Danish Academy of Fine Arts (2011-12), Rice University (2014-15), Delf University of Technology (2015-16), and Columbia University (2017), General Incorporated Association, "Small Planet" Director (2020). Main works: see the list of Kaijima's. Main books: "Atelier Bow-Wow Graphic Anatomy" (2007, TOTO Publishing), "Window Scape" (2010, Film Art), etc.

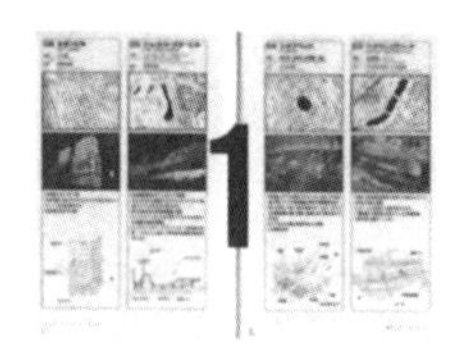

BIBLIOGRAPHY
参考书目

1996/12/10 『磯崎新の革命遊戯』p.223–280 TOTO 出版 [1]
「ダメ建築礼賛」p.249–271
「メイド・イン・トーキョー」p.249–271
「空白恐怖症の東京」p.272–280
Arata Isozaki's Revolution Game p.223–280, TOTO Publishers
Dame-Architecture p.249–271
made in tokyo p.249–271
Void-Fear in Tokyo p.272–280

1997/01 『新建築』9701/p.195–198 [2]
特別記事：96 アーキテクチュア・オブ・ザ・イヤー批評
「カメラ・オブスキュラあるいは革命の建築博物館」展
すなわち磯崎新のエンドレス・デスマッチ　土居義岳
Shinkenchiku 9701 p.195–198
Architectural Museum of Revolution Exhibition
or Arata Isozaki's Endless Match by Yoshitake Dohi

1997/02 『建築文化』vol.52/no.604/1997.2/p.10 [3]
「建築史とのつきあい方」橋本憲一郎
Kenchikubunka vol.52 no.604 1997, 2 p.10
How to play with Architecture History by Kenichiro Hashimoto

1997/04/01 『Inter Communication』20/p.8–9/NTT 出版 [4]
「都市の未来形型」メイド・イン・トーキョー
Inter Communication 20 p.8–9 NTT publishing
The Shape of Cities in the Future [I]

1998/02/10 『10+1』No.12/p.80–90/INAX 出版 [5]
「他者が欲望する黒船都市，トーキョー」五十嵐太郎
“10+1” No.12, p.80–90, INAX
Delirious blackship city, Tokyo by Taro Igarashi

1998/04/01 『SD』9804/p.52/ アトリエ・ワン Recent Works/ 鹿島出版会 [6]
SD 9804 p.52 Atelier Bow-wow Recent Works Kajima Publishing

1998/07/25 『A』1998 vol.1/PULP CITY/p.20–21 [7]
「メイド・イン・トーキョー」=New Urban Architecture Style
A 1998 vol.1, PULP CITY, p.20–21
Made in Tokyo=New Urban Architecture Style

1998/07/30 『山形新聞』7 面，うえーぶ 98「東京の“ダメ建築”集める」[8]
Yamagata Shinbun P7 WAVE 98
“Collecting Dame-Architecture in Tokyo”

1998/08/24 『日経アーキテクチュア』p.102–105/ 日経 BP 社 [9]
「都市の再発見 City Observation」
Nikkei Architecture p.102–105 Nikkei BP
“Re-discovering the City: City Observation”

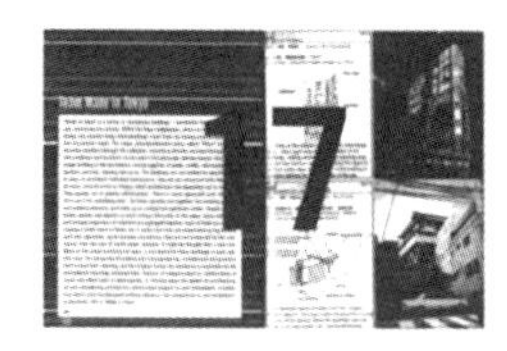

1998/08/29 『Zuritip』p.68「Made In Tokyo」[10]

1998/10/10 『20 世紀建築研究』p.260-279/「B 級 BT」INAX 出版
"20th Century Architecture Reserch" p.260-279
"B Grade Building Type" INAX Publishers

1999/01/20 『読売新聞』夕刊 4 面 /「複雑系」都市 東京の魅力 [11]
Yomiuri Shinbun Evening News Paper p.4
"Complex City: the Fascination of Tokyo"

1999 『Techniques』p.82 p.218 UN Studio [12]

1999/05 『HIROBA』表紙 / 近畿建築士会協議会 [13]
「デジタルアーキテクチュア 29」
Hiroba front page, Kinkikenchikushikai-kyougikai
"Digital Architecture 29"

1999/05 『archis』p.42-43「Made in Tokyo」[14]

1999/06/01 『AXIS』vol.79/p.26-29「Made in Tokyo」
株式会社アクシス [15]
AXIS vol.79 p.26-29 Made in Tokyo AXIS

1999/07 『建築雑誌』vol.114/no.1441/1999.7/p.22 AIJ
「トーキョー・ガイドブック」
Architecture Magazine vol.114 no.1441 1999.7 p.22
"Tokyo Guide Book" AIJ

1999/10/15 『CREATORS'FILE for LIVING-
ニッポンのクリエーター 58 人のしごと』, p.103, ギャップ出版 [16]
MADE IN TOKYO [メイド・イン・トーキョー]
"CREATORS'FILE_for LIVING" p.103,
Gap Publishing "MADE IN TOKYO"

1999/12/01 『建築キーワード』p.409-411、住まいの図書館
Architectural Keyword dictionary p.409-411

2000 『ベネツィアビエンナーレ 2000』「都市」http://www.labiennale.org/
Venice Biennale 2000 City http://www.labiennale.org/

2000 『la Biennale di Venezia on line』
「Team Made in Tokyo」p.336-339 [17]

2000/6 『ARCH+』151 p.52-53「Made in Tokyo」[18]

2000/11 『SiA』Nr.44「MIT = Made In Tokyo」p.27-33 [19]

2001/3 『Limits』Col.legi. d'Arquitectes de Catalunya
「MADE IN TOKYO」p.4 [20]

图书在版编目（CIP）数据

东京制造 / (日) 贝岛桃代 , (日) 黑田润三 , (日) 塚本由晴著 ;
林煌译 . -- 北京 : 北京联合出版公司 , 2023.9
ISBN 978-7-5596-7003-8

Ⅰ . ①东… Ⅱ . ①贝… ②黑… ③塚… ④林… Ⅲ .
①建筑艺术－研究－东京 Ⅳ . ① TU-863.13

中国国家版本馆 CIP 数据核字 (2023) 第 108462 号

北京市版权局著作权合同登记号 图字：01-2023-3552 号
审图号：GS（2023）1809 号

东京制造
作　　者：［日］贝岛桃代 黑田润三 塚本由晴
译　　者：林　煌
出 品 人：赵红仕
策划机构：明　室
策 划 人：陈希颖
特约编辑：李佳晟　王佳丽
责任编辑：高霁月
装帧设计：山川制本 workshop
原版设计：古平正义
照片提供：Takashi Homma 金町驾校 东京制造团队

Authors：Momoyo Kaijima, Junzo Kuroda, Yoshiharu Tsukamoto
Design(Japanese edition)：Masayoshi Kodaira
Photo：Takashi Homma（cover, endpaper, P0-1, P34-35）
Kanamachi Driving School（P93）
Team Made in Tokyo

北京联合出版公司出版
（北京市西城区德外大街 83 号楼 9 层　100088）
北京联合天畅文化传播公司发行
北京市十月印刷有限公司印刷　新华书店经销
字数 134 千字　880 毫米 ×1230 毫米　1/32　6 印张
2023 年 9 月第 1 版　2023 年 9 月第 1 次印刷
ISBN 978-7-5596-7003-8
定价：52.00 元